STORIES IN STONE

Teacher's Guide

Grades 4–9

Skills
Observing, Comparing, Identifying Properties, Classifying,
Forming Hypotheses, Conducting Experiments,
Drawing Conclusions, Inferring, Using Simulations

Concepts
Rocks and Minerals, Applying Standard Methods of Classification,
Sedimentary, Igneous, and Metamorphic Rocks, Erosion and Deposition,
Melting and Crystallization, Crystal Shapes, Earth Processes, The Rock Cycle

Themes
Patterns of Change (Cycles), Scale, Structure, Systems and Interactions,
Diversity and Unity, Models and Simulations, Evolution
(For more on Themes in GEMS, please see page 156.)

Math Strands
Patterns, Geometry, Logic and Language

Time
Eight 40- to 60-minute sessions

by
Kevin Cuff
with **Cary Sneider, Lincoln Bergman,**
Alan Gould, JohnMichael Seltzer

LHS GEMS

Great Explorations in Math and Science (GEMS)
Lawrence Hall of Science
University of California at Berkeley

Cover Design
Rose Craig

Photographs
Richard Hoyt
Laurence Bradley

Illustrations
Carol Bevilacqua
Rose Craig
Lisa Klofkorn

Lawrence Hall of Science, University of California,
Berkeley, CA 94720

Chairman: Glenn T. Seaborg
Director: Marian C. Diamond

Initial support for the origination and publication of the GEMS series was provided by the A.W. Mellon Foundation and the Carnegie Corporation of New York. Under a grant from the National Science Foundation, GEMS Leader's Workshops have been held across the country. GEMS has also received support from the McDonnell-Douglas Foundation and the McDonnell-Douglas Employees Community Fund, the Hewlett Packard Company, Join Hands Educational Alliance, and the people at Chevron USA. GEMS also gratefully acknowledges the contribution of word processing equipment from Apple Computer, Inc. This support does not imply responsibility for statements or views expressed in publications of the GEMS program. For further information on GEMS leadership opportunities, or to receive a catalog and the *GEMS Network News*, please contact GEMS at the address and phone number below. We also welcome letters to the *GEMS Network News*.

International Standard Book Number: 0-912511-93-1

COMMENTS WELCOME !

Great Explorations in Math and Science (GEMS) is an ongoing curriculum development project. GEMS guides are revised periodically, to incorporate teacher comments and new approaches. We welcome your criticisms, suggestions, helpful hints, and any anecdotes about your experience presenting GEMS activities. Your suggestions will be reviewed each time a GEMS guide is revised. Please send your comments to: GEMS Revisions, c/o Lawrence Hall of Science, University of California, Berkeley, CA 94720.
The phone number is (510) 642-7771

Great Explorations in Math and Science (GEMS) Program

The Lawrence Hall of Science (LHS) is a public science center on the University of California at Berkeley campus. LHS offers a full program of activities for the public, including workshops and classes, exhibits, films, lectures, and special events. LHS is also a center for teacher education and curriculum research and development.

Over the years, LHS staff have developed a multitude of activities, assembly programs, classes, and interactive exhibits. These programs have proven to be successful at the Hall and should be useful to schools, other science centers, museums, and community groups. A number of these guided-discovery activities have been published under the Great Explorations in Math and Science (GEMS) title, after an extensive refinement process that includes classroom testing of trial versions, modifications to ensure the use of easy-to-obtain materials, and carefully written and edited step-by-step instructions and background information to allow presentation by teachers without special background in mathematics or science.

Staff
Glenn T. Seaborg, **Principal Investigator**
Jacqueline Barber, **Director**
Kimi Hosoume, **Assistant Director**
Cary Sneider, **Science Curriculum Specialist**
Jaine Kopp, **Mathematics Curriculum Specialist**
Carolyn Willard, **GEMS Centers Coordinator**
Laura Tucker, **GEMS Workshop Coordinator**
Katharine Barrett, Kevin Beals, Ellen Blinderman,
Beatrice Boffen, Gigi Dornfest, John Erickson,
Linda Lipner, Laura Lowell, Debra Sutter,
Rebecca Tilley, **Staff Development Specialists**
Jan M. Goodman, **Mathematics Consultant**
Cynthia Eaton, **Administrative Coordinator**
Karen Milligan, **Distribution Coordinator**
Terry Cort, **Workshop Administrator**
Felicia Roston, **Shipping Coordinator**
Stephanie Van Meter, **Trial Testing Coordinator**
Lisa Haderlie Baker, **Art Director**
Carol Bevilacqua, Rose Craig, Lisa Klofkorn, **Designers**
Gerri Ginsburg, **Public Information Representative**
Lincoln Bergman, **Principal Editor**
Carl Babcock, **Senior Editor**
Florence Stone, **Assistant Editor**
Kay Fairwell, **Principal Publications Coordinator**
Larry Gates, Lisa Ghahraman, George Kasarjian,
Alisa Sramala, Mary Yang, **Staff Assistants**

Contributing Authors

Jacqueline Barber	Alan Gould
Katharine Barrett	Kimi Hosoume
Kevin Beals	Susan Jagoda
Lincoln Bergman	Jaine Kopp
Beverly Braxton	Linda Lipner
Kevin Cuff	Laura Lowell
Celia Cuomo	Larry Malone
Linda De Lucchi	Cary I. Sneider
Gigi Dornfest	Craig Strang
Jean Echols	Debra Sutter
Philip Gonsalves	Rebecca Tilley
Jan M. Goodman	Jennifer Meux White
	Carolyn Willard

Reviewers

We would like to thank the following educatorswho reviewed, tested, or coordinated the reviewing of *this series* of GEMS materials in manuscript and draft form (including the GEMS guides *Stories in Stone* and *Math Around the World*). Their critical comments and recommendations, **based on classroom presentation of these activities nationwide**, contributed significantly to these GEMS publications. Their participation in the review process does not necessarily imply endorsement of the GEMS program or responsibility for statements or views expressed. Their role is an invaluable one, and their feedback is carefully recorded and integrated as appropriate into the publications. **THANK YOU!**

ALASKA
Wasilla Middle School, Wasilla
Cynthia Dolmas Curran *
Michael Curran
Suzanne Cyr
Fred Hajduk
Gary Walker

ARIZONA
Northern Arizona University, Flagstaff
Lynda Hatch **
Andrea Temple

Sechrist Elementary School, Flagstaff
Avis Berg
Penny Bettes
Diane Eide
Deb Gaither *

Esperanza Elementary School, Tucson
Katie Dawson *
Carol Duggan
Marla Finn
Hans Schot

Topawa Middle School, Topawa
Charlet Connelly
Beverly Hawley
Margo Owen
Chris Pray *

CALIFORNIA
Fresno
Forkner Elementary School
Wanda Begley
Bruce Lundberg
Grant Phillips
Cindy Schnell *

Los Angeles
Foshay Learning Center
Carolyn Brown
Corrine Kimmell
Karel Lilly *
Christine Yap

Westside Science Center
Nonnie Korten **

San Francisco Bay Area
Bancroft Middle School, San Leandro
Catherine Heck
Paul Hynds *
Julie Privitera

Calvin Simmons Junior High, Oakland
Stan Lake *
William Paul
Emiliano Sanchez

Downer Elementary School, San Pablo
Lourdes Gonzales
Linda Hanna
Nancy Hirota
Lola Taylor

Everett Middle School, San Francisco
Julio Burroughs
Margo Commando *
Nichole Gottfried
Mildred Harkless-Webb
Adele Lawrence

Havenscourt Junior High, Oakland
Roy Caldwell
Wajibu Green *
Vincent Tolliver
Margaret Velasquez

Joaquin Miller Elementary School, Oakland
Ben Lang *
Kathy Lee
Joyce Milten

Lazear Elementary School, Oakland
Carla Anders
Lorna Baird
Jose Davila
Sophia Estrella *

Longfellow School, Berkeley
Tyrdah Alafia-Young
Crispin Barrere *
Caroline Chun

Malcolm X Intermediate, Berkeley
Mahalia Ryba *
DeEtte LaRue
Jessie Shohara
Gloria E. Thornton

Marie A. Murphy Elementary, Richmond
Patsy Christner
Debby Kerreos-Callahan
Mary Tutass
Sandra A. Petzoldt *

Martin Luther King, Jr. Junior High, Berkeley
Mark Delepine
Peter Levitt
Beth Sonnenberg
Ada Wada *

Montera Junior High, Oakland
Russ Bettoncort
Mary Charlesworth
Alice Holmes
Matt Smith

Saint Perpetua's School, Lafayette
Eileen Cafferty
Susan Falaschi
Carol Grover *
Sister Theresa

Stanley Intermediate School, Lafayette
Sherry Cutting
Sheila Geritz
Glenn Hoxie *
Laurie Rader

Washington Elementary, Oakland
Vince Avila

Westlake Junior High, Oakland
Philip Makau, Ph. D. *
Carlos Pineda

Whittier Year Round Elementary, Oakland
Pamela Booker
Willie M. Crawford *
Debra Dornier-Arrizón
Amy Introligator
Linda L. Wrice

Wilson Elementary School, San Leandro
Jim Bolar *
Julie Chung
Jenny Drake
Noreen Ford
Mari Patton

KENTUCKY
Brown School, Louisville
Robin Lipsey
Bill Munro-Leighton
Rita Nichols
Tony Peake *

MAINE
Bowdoinham Community School, Bowdoinham
David Galin *
Thomas Millay
Luana Smith
Kathleen Stehle

MASSACHUSETTS
McCarthy Middle School, Chelmsford
Marsha Kelly
Nancy Kiernan

Parker Middle School, Chelmsford
Donna Foley *
Mark Johnson
Judy White

MICHIGAN
Coburn Elementary, Battle Creek
Betty Hansen
Anna Horvatinovich
Bev Osborn
Tim Schillaci *
Diana Smyth
Jackie Zanotti

St. Clair ISD, Port Huron
Carl Arko **

NEW YORK
Buffalo City Schools
Stanley J. Wegrzynowski ** (state coordinator)

Frederick Law Olmsted Public School #56, Buffalo
Virginia Coon
Tim Maul
Francine R. Shea *
Kirby Sneider

Isaac E. Young Middle School, New Rochelle
Joan DeSantolo
Vincent D. Iacovelli *
Nona Johnson
Emily Reid
Regina Simoes

NORTH CAROLINA
University of North Carolina at Chapel Hill
MSEN Pre-College, Chapel Hill
Eric Packenham **

Sheppard Middle School, Durham
Sherline Hankins
Retella Jones *
Shola Kujore

C. W. Stanford Middle School, Hillsborough
Leslie K. Jones *
Connie Lancaster
Beth Neill
April Torrington

PENNSYLVANIA
Cedar Crest Middle School, Lebanon
Margaret Beatty
Michael Doll
Camilla Marozin *
Rebecca Snader

TEXAS
Clear Creek ISD, League City
Julia Haun **

Wedgewood Elementary, Friendswood
Carol Fernandez *
Barbara Holton
Charles Meredith
Trish Tagliabue

WASHINGTON
Challenger Elementary School, Issaquah
Roberta Andresen *
Diane Fielding
Susan Rogers
John Sage

WEST VIRGINIA
Sissonville Middle School, Sissonville
Vicky Carney
Lisa Elliott
Margaret Miller *
Beth Porter

Sunrise Museum, Charleston
Andrea Ambrose **

*   On-Site Coordinator
** Regional Coordinator

Contents

Acknowledgments

The source for some of these activities derives from numerous after-school classes presented at the Lawrence Hall of Science (LHS). In particular, the salol simulations were derived, modified, and refined from activities in a class entitled "Spaceship Earth," initially developed by Cheryl Jaworowski, then a graduate student in geology at the University of California, Berkeley. Kevin Cuff, the main author of this guide, a geologist by training, took the lead in adapting these salol activities for this guide, and in originating, developing and/or recasting the other main activities for *Stories in Stone*. He was assisted by Cary Sneider, Alan Gould, JohnMichael Seltzer, and GEMS Principal Editor Lincoln Bergman. David Siegel, Frank Trusdell, and Kaunda Collingwood all offered valuable suggestions to the main author concerning earlier versions of this guide. Further suggestions were provided in the formative and testing stages of this guide by GEMS Director Jacqueline Barber, along with GEMS educators Carolyn Willard and Laura Tucker. The search for "Literature Connections" was assisted by GEMS Assistant Editor Florence Stone. Special thanks is also extended to GEMS staff members Cynthia Eaton, for arranging GEMS field trials, and Stephanie Van Meter, for putting together the classroom kits used in trial testing.

The idea of extending Session 3 to enable students to learn about Euler's Theorem was suggested by Margaret Smart, Bancroft Middle School, San Leandro, California. Harriette Stevens, Co-Director of the ACCESS Project at LHS, provided further information on Euler's Theorem. Thanks to Paul Hynds and the science club at Bancroft Middle School for letting us take some of the photographs used in this guide.

We are particularly appreciative of the efforts of all the teachers who helped in the local and national testing of this guide (listed near the front of this book) and most especially their students. Many students took the time to send us their own reactions to the unit and their questions prompted by the activities. Several classes sent us colorful mobiles made from the three-dimensional crystal models of Session 3. Roberta Andreson's class at Challenger Elementary in Issaquah Washington sent us some of their imaginative drawings and related work, including several that appear in this guide.

The color photographs of the ten class samples were obtained from and used with the permission of Ward's Natural Science Establishment in Rochester, New York. We appreciate the use of these photos, which appear with numerous others in Ward's catalog. The address and phone number of Ward's appears in the "Resources" section of this guide. The photographs of "lava rivers" in front of the "Behind the Scenes" section and on the inside back cover were taken by Kevin Cuff.

We would also like to thank Grove Press, Inc., for granting permission to reproduce excerpts from the poetry of Pablo Neruda. The excerpts are from *Pablo Neruda Five Decades: A Selection (Poems: 1925–1970)*, a bilingual edition edited and translated from the Spanish by Ben Belitt, copyright 1974 by Grove Press, Inc. The Spanish poems are also copyright by the Pablo Neruda Foundation and the English translations are also copyright by Ben Belitt. This poetry was chosen because it so beautifully exemplifies the processes students explore in this unit. These and many other poems of Neruda make wonderful world and Latin American literature connections for all students. Your Spanish-speaking students may especially enjoy learning more about these poems and the poet. This Grove Press edition is widely available at bookstores and libraries, and contains the full text of Neruda's original Spanish, as well as fine translations.

Introduction

Just about everyone loves rocks. We are drawn to their color and texture, shape and contour. There is a special attraction and power in them, be it felt in childhood rock collections, lucky charms, beautifully-fashioned jewelry, or the sacred stones of many cultures. They are the stuff of which the crust of our precious home planet Earth is made.

The clay pottery and chiseled statues created by artists of all ages and cultures are transformed from the Earth. Countless everyday materials and much technology depends on what humans derive from mining, refining, shaping, combining, and transforming substances derived from rocks and minerals. The soil within which plants that sustain life on Earth grow, is mostly made of rock fragments and minerals.

So many epic stories are hidden in stone— fossils of the prehistoric past, clues to geologic, environmental and weather conditions—the record of centuries, of eons. Over time, humans have learned to read some of these stories, whose powerful events may range from immense volcanic cataclysm to slow steady layering, grain by grain.

*Geologists study all aspects of Earth systems. The study of rocks, called **petrology**, is a subfield of geology.*

It is important to convey to students that the Earth Sciences are far from static—they are concerned with constant processes of change. Theories about the Earth's shifting tectonic plates have transformed the Earth Sciences in the past few decades. While rocks and minerals are often regarded as among the most static and solid objects we know, the body of knowledge of which they are a part—like all science—is always changing as we learn more, modify, and transform our understandings.

By examining actual specimens of the Earth's crust, students learn about basic processes that have shaped and transformed the Earth over billions of years. Student experiences with these real objects, rocks and minerals, combined with simulations of Earth processes, lead to deeper understandings and more abstract concepts, such as how rocks form, rock cycles, and plate tectonics.

Throughout the unit, students find and collect interesting rocks and minerals, observe them with magnifying lenses and compare different samples, and make inferences about large classes of rocks—how they change and are related to each other. In addition, the students conduct simulations and an experiment to find out more about how rocks and minerals are formed.

Session-By-Session

In Session 1, students examine a class collection of rock and mineral samples. The focus is on observing and appreciating the properties of rocks and minerals, as well as evoking curiosity about how different kinds of rocks and minerals might be formed. The samples are numbered and students receive a key, giving the names of the rocks and minerals from the start. We recommend that you refer to the samples by their numbers at first, so that memorizing names does not get in the way of students observing and devising their own classification schemes. Soon, many students will notice the key and start referring to the rocks and minerals by their names. When that happens you can make an easy transition to using the names rather than (or in addition to) the numbers, but the primary emphasis in the unit is on properties, processes, and the students' own observations and experiences, not nomenclature.

As in other GEMS units, the standard method for classifying rocks and minerals is not introduced until the students have had an opportunity to investigate the samples and identify the properties that *they* think may be important in distinguishing one sample from another. It's important not to short-circuit this process, as it both empowers and challenges the students, allowing them to exercise all their powers of free observation and creative classification.

Later, when the standard methods and terms are introduced, the students will have gained their own insights into the skills, methods, and thought processes other Earth scientists must utilize when they classify rocks and minerals. At the end of Session 1, and on subsequent days, students are asked to look for and bring in their own "mystery rocks" for use later in the unit and perhaps get them started making their own rock collections.

Special Note to Teachers Who Are "Rockhounds" and Collectors

You will note that this unit does not stress some of the other methods of testing/ classifying rocks, using such devices as a streak plate, or items for assigning a numerical value to hardness. The emphasis of Stories in Stone *is on observation of properties that lead to understanding of the **processes** that form different rocks and minerals, and on what this can tell us about the changing crust of our planet. If you would like to introduce your students to the many other techniques used to identify rock and mineral specimens, we encourage you to do so after their experience with this unit. The more specific tests and information should make more sense within the broader framework for understanding rock and mineral formation provided by this guide.*

In Session 2, students are introduced to the distinction between a rock and a mineral. They create mineral crystals of sodium chloride, or salt, which helps them connect minerals and rocks with an everyday household substance. The cubic structure of these salt crystals is also the same as that seen on the halite found in their class samples. An optional "Minerals at Home" handout t is included after this session, with an activity suggestion to help students realize how important minerals are in their daily lives. This can be used at any point in the unit.

Session 3 features an engaging activity in which students fold paper templates to make models of crystalline shapes. This connects to the crystals they've observed, intersects strongly with geometry and topology, and could make a nice introduction to a deeper study of crystallography.

In the following sessions, students become involved in a series of hands-on activities that focus on the processes that are responsible for the formation of rocks and minerals. Students, working together, experiment, model and simulate to gain insight into three basic processes through their own hands and minds. They gain initial understandings of how igneous, metamorphic, and sedimentary rocks are formed and learn that these three main types provide the basis for one of the standard classification systems that Earth scientists use to classify rocks and minerals.

In Session 4, Formation of Igneous Rocks, students conduct an experiment that simulates how igneous rocks form when molten material (magma) cools and solidifies. They use phenyl salicylate (salol) to create a molten material, then observe the formation of crystals at two different temperatures. Based on their findings, especially concerning how the size of the crystals relates to temperature, they discuss three igneous rock samples from the class collection.

As with all experiments, the results can be unpredictable, and some student groups may not come to the same conclusions as others. This is a great opportunity for students to discuss and draw their own unique conclusions about why such variation might take place, and how they (and other scientists) might deal with situations in which similar experiments come out differently, or in which the same results are interpreted differently by different people. Be sure to provide ample time for discussion. Explain that controversy is part of the process by which science grows and changes. By all means, avoid telling the students that they have done the experiments "wrong" if their results are not what you expected. If students don't suggest it themselves, suggest re-doing experiments or elicit student suggestions for additional experiments that might yield more conclusive results.

In Session 5, Formation of Sedimentary Rocks, students investigate sediments of different grain sizes—sand, silt, and clay. They suspend a mixture of these materials in water and allow them to settle, gaining a sense of the layering process so key to the formation of sedimentary rocks. In the same way, they obtain and investigate soil samples from near the school, comparing how they settle and layer with the first settling process, and estimating the relative proportions of sand, silt, and clay in local soil.

Session 6, Formation of Metamorphic Rocks, models the processes that produce the intense heat and pressure which create metamorphic rocks. Students create clay models as the teacher tells an environmental "rock formation" story that establishes the conditions for the changes that need to take place for metamorphic rocks to form.

In Session 7, Recycling the Earth's Crust, students continue using clay to actively model the rock cycle—the idea that, over the course of time, one type of rock can be transformed into another. This important but abstract concept is thus "grounded," so to speak, in the students' own hands. This "story in stone" narrative builds on what students have learned about properties and processes to foster further understanding of how the rocks and minerals of the Earth's crust are subjected to changes caused by geological and environmental events. This of necessity involves geological events that can best be understood through the modern theory of plate tectonics. For younger students, the basic idea that the Earth's crust is always moving may be sufficient, while older students can be encouraged to delve more deeply into the fascinating idea that the Earth's crust is composed of large, shifting plates.

Full many a gem of purest ray serene,
The dark unfathomed caves of ocean bear…
        Thomas Gray
        Elegy Written in a
        Country Churchyard

In Session 8, students bring their growing knowledge to bear on drawing their own conclusions about the samples in the class sets. Using the student data sheet provided, they classify all the samples as to whether or not they are rocks or minerals, and, if they are rocks, whether they are igneous, sedimentary, or metamorphic. Student groups then go on to face the challenge of classifying one of the "mystery rocks." A "Going Further" language arts activity features poem-riddles for each rock that can be matched by students to each of their samples. Students could write a similar poem for a "mystery rock." Many other rich writing opportunities occur throughout the guide.

We hope these activities succeed in introducing you and your students to the multi-layered stories revealed by investigation of rocks and minerals. In this unit, your students have a chance to experience, through hands-on activities, simulations of important processes that give rise to and transform these pieces of the Earth's crust we call rocks and minerals. In doing so, they gain an intuitive, experiential understanding of some of the motive forces constantly at work—as the Earth's plates shift, as weathering and erosion take their course. In this way, students become better able to construct their own understandings of the "stories in stone."

Take a stroll outdoors. As you walk, look along the ground for a rock of your choice. Pick it up. You may want to wash it off to help bring out any deep underlying colors, textures, lines, and layers. Bring it to the classroom and put it on your desk. This can be your "mystery rock."

The Storytelling Stone

A Seneca Indian story tells where the first stories came from. During a long cold winter, a boy out hunting pauses to rest near a great rock shaped almost like the head of a person. He hears a deep voice. "The Great Stone," (who the boy calls Grandfather) tells him the very first story, about the creation of the Earth. The boy goes back to tell the people, and the story seems to drive away the cold and help the people make it through the winter. The boy goes back again and again, with offerings for "The Great Stone," and learns more stories, until the time the large rock tells him that there are no more. "I have told you all my stories...Now the stories are yours to keep for the people. You will pass these stories on to your children and other stories will be added to them as years pass. Where there are stories, there will be more stories. I have spoken..."

You may want to read this story to your class or have them read it themselves. It appears at the beginning of *Keepers of the Earth: Native American Stories and Environmental Activities for Children* by Michael J. Caduto and Joseph Bruchac, Fulcrum, Inc., Golden, Colorado, 1988. For this unit, we might respectfully build upon this story to include the idea that there are still many stories in the stone, and it is up to people, for example, your students, to learn more about these stories for themselves, as well as to create their own!

Time Frame

The preparation times estimated below include time to set up the classroom and prepare special materials. Classroom activity time will vary greatly depending on your particular group of students, as well as your own needs and preferences.

Session 1:   **Properties of Rocks and Minerals**
Preparation   60–90 minutes
Class Activity   40–60 minutes

Session 2:   **Distinguishing Rocks from Minerals**
Preparation   20–30 minutes
Class Activity   40–60 minutes

Session 3:   **The Shapes of Mineral Crystals**
Preparation   20–30 minutes
Class Activity   40–60 minutes

Session 4:   **Formation of Igneous Rocks**
Preparation   60–90 minutes
Class Activity   40–60 minutes

Session 5:   **Formation of Sedimentary Rocks**
Preparation   30–40 minutes
Class Activity   40–60 minutes

Session 6:   **Formation of Metamorphic Rocks**
Preparation   30–40 minutes
Class Activity   40–60 minutes

Session 7:   **Recycling the Earth's Crust**
Preparation   15–20 minutes
Class Activity   40–60 minutes

Session 8:   **Classifying Rocks and Minerals**
Preparation   15–20 minutes
Class Activity   40–60 minutes

Creating a Class Kit for Stories in Stone

Obtaining Sets of Rocks and Minerals

It is impossible for students to learn about rocks and minerals without rocks and minerals! To keep costs at a minimum, we have designed the unit around sets of ten common rocks and minerals that can be obtained from a variety of different sources. Once you obtain enough sets for your class to do these activities, plus the materials listed on the next couple of pages, you will have a *Stories in Stone* Class Kit to use in future years, or for sharing with other teachers.

You will need one set of rocks and minerals for the teacher, and one for each lab team in your class. We strongly recommend lab teams of no more than four students for these activities. So, for a class of 32 students, you will need nine sets of the following rocks and minerals.

1. schist	6. galena
2. halite	7. slate
3. granite	8. obsidian
4. quartz	9. shale
5. basalt	10. conglomerate

All samples should be at least thumbnail size—but the larger the better. It is ideal for the teacher's set to have large samples, so when the teacher holds up a sample it is easier for students to see it. The ten rocks and minerals on the list have been carefully chosen to represent major rock types and important minerals, and yet be commonly available in most areas of the United States. We recommend you obtain extra samples in case some are lost or some samples are damaged. You can obtain them from any of several sources, for example:

Commercial science mail-order companies sell rock samples for classroom use. In the margin are the names of two companies from which you can buy all of the rocks and minerals you will need. When you place your order, we recommend that you also order an ounce or two of **salol crystals** (phenyl salicylate), that will be needed for Session 4. See the "Resources" section for books and other educational resources that may be helpful.

Ward's Natural Science Establishment., Inc.
5100 W. Henrietta Rd.
P.O. 92912
Rochester, NY 14692-9012
(800) 962-2660

Frey Scientific
905 Hickory Lane
P.O. Box 8101
Mansfield, OH 44901-8101
(800) 225-FREY

Local rock shops can often be found in the "Yellow Pages" of your phone book. Although the list of ten rocks and minerals is ideal for teaching this unit, if a sample is not available you can substitute rocks or minerals of the same type. However, if you do make a substitution, you will need to change the keys to the samples accordingly.

Parents, hobbyists, and local "rock hound" clubs may be able to provide you with samples. Send a note home to your students' parents with a list of the rocks and minerals you will need and ask for help or referrals to others who might help.

Preparing Class Sets of 10 Rocks and Minerals

1. Separate all samples of the same type into ten small piles and familiarize yourself with their characteristics and names. Next place a small—about 1/4 inch (0.5 centimeters) in diameter—dab of white correction fluid on the smoothest surface of each specimen. Allow at least two hours for drying.

2. After the correction fluid is completely dry, use a fine line permanent ink marker to number the white spots on individual samples as follows:

1. schist	6. galena
2. halite	7. slate
3. granite	8. obsidian
4. quartz	9. shale
5. basalt	10. conglomerate

3. To ensure that the numbers do not rub off, you may want to coat each number with a dab of clear nail polish.

4. Prepare one "Rocks and Minerals Key" for each set, by copying the master on page 20 and cutting it into six pieces.

5. Create sets of rocks and minerals by placing one of each sample into an egg carton. Put four magnifying lenses into the empty compartments, and tape a copy of the "Rocks and Minerals Key" to the inside of the lid.

Materials Needed for All Sessions

On these four pages is a complete list of all materials that you will need to teach *Stories in Stone* for a class of 32 students. You may want to assemble them in one or two cardboard "banker's" boxes, or in large plastic containers, and label them: "Stories In Stone—Class Kit." (Note: The terms "consumable" and "non-consumable" are used to distinguish between items that will or will not be used up as you present the unit. The rock samples, for example, are non-consumable—they can be used for many classes; the salt or salol are consumable items and must be replenished. A third category, general supplies, includes items such as pencils and paper.)

Non-Consumables for *Stories in Stone*

Rock Samples

- ❏ 9 samples of the following rocks and minerals*
   (plus one or two extra samples)

schist	halite	granite
quartz	basalt	galena
slate	obsidian	shale
		conglomerate

- ❏ 9 egg cartons
- ❏ 32 magnifying lenses
- ❏ 1 bottle of white correction fluid, to label rocks
- ❏ 1 bottle of clear nail polish, to label rocks
- ❏ 2 copies of "Rocks and Minerals Key," page 20
- ❏ 8 trays
- ❏ 16 sturdy paper plates
- ❏ 1 coffee pot, hot plate, or other source of hot water
- ❏ 24 small plastic cups (about 2 oz. capacity)
- ❏ 16 clear or black plastic spoons
- ❏ 32 stir sticks or additional plastic spoons
- ❏ 9 votive candles with holders
- ❏ 18 metal spoons (34 spoons are ideal)
- ❏ 1 quarter-teaspoon measuring spoon
- ❏ 32 pairs of goggles
- ❏ 16 clear plastic 6–10 oz. plastic cups
- ❏ 1 pitcher with 2–3 quarts of water
- ❏ 2 sponges
- ❏ 1 large bucket for waste water and soil
- ❏ 8 plastic knives for cutting clay
- ❏ 8 "mystery rocks"

Optional
- ❏   8 flashlights

* Order from scientific supply companies.

Correction Fluid

Small Plastic Cups

Large Plastic Cups

Measuring Spoon

Clear Nail Polish

Votive Candles

Plastic Spoons

Goggles

Magnifying Lenses

Paper Plates

Trays

Plastic Knives

Stir Sticks

Metal Spoons

Sponges

Rocks and Minerals Key

Pitcher

Bucket

Egg Cartons

Coffee Pot

Consumables for *Stories in Stone*

- ❏ 1 cup of Kosher salt (table salt can be substituted)
- ❏ 1 book of matches
- ❏ ice cubes
- ❏ 1 container of salol crystals (2 oz. is adequate for a class)*
- ❏ 1 pound each of sand, silt, and clay (in plastic bags or jars)
- ❏ 1 container of modeling clay of the following colors:
  (note: 8 bars modeling clay = 1 pound)
  red   4 bars   green   8 bars
  yellow 4 bars   blue   12 bars
- ❏ 33 copies of each of the following sheets:
  "Cube," page 40
  "Hexagonal Prism and Pyramid," page 41
  "Observing Crystal Formation," page 62 (or 63)
  "Rock Type Description and Classification," page 110
  "Observation and Display of Mystery Rock," page 111

* Order from scientific supply companies.

Kosher Salt

Matches

General Supplies for *Stories in Stone*

- ❏ 32 pencils
- ❏ 8 sheets of blank paper
- ❏ 8 index cards
- ❏ 32 pairs of scissors
- ❏ 16 rolls of transparent tape
- ❏ 1 roll of masking tape
- ❏ 1 stack of newspapers to cover student work areas
- ❏ 1 fine line permanent marker, to label rocks
- ❏ 1 heavy line permanent marker, to label bags of soil
- ❏ 16 white paper towels

Ice Cubes

Salol

Clay

Silt

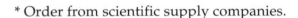

Sand

Red Clay

Yellow Clay

Green Clay

Blue Clay

Pencils

Blank Paper

Index Cards

White Paper Towels

Cube

Hexagonal Prism & Pyramid

Observing Crystal Formation

Rock Type Classification Sheet

Observation and Display of Mystery Rock

Student Data Sheets

Scissors

Masking Tape

Fine Line Marker

Heavy Line Marker

Newspaper

The Daily Sediment

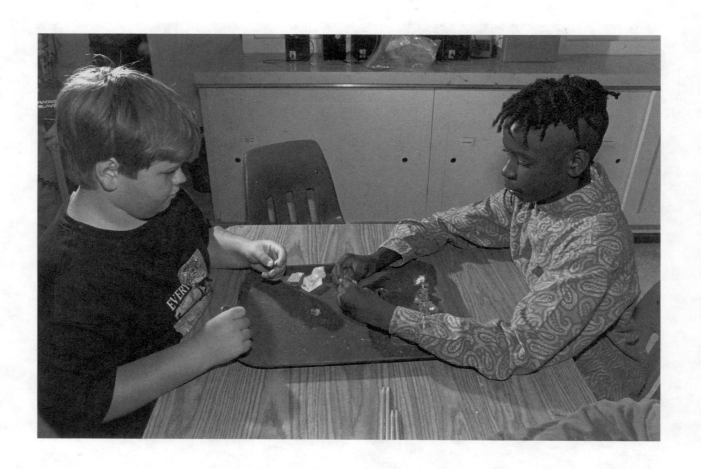

Session 1: Properties of Rocks and Minerals

Our lives—and the lives of all plants and animals—begin, evolve, and end on the outermost layer of the Earth known as the crust. To gain a better understanding of what the Earth's crust is made of, students work in groups to carefully observe and describe important properties of ten samples of crustal material—ten samples of rocks and minerals.

In this session, student groups look for properties—observable characteristics—that can be used to distinguish one sample from another. Students arrange their samples into different groups or clusters, according to the property that they have selected. When all student groups have completed their arrangements, each group displays their results while others are invited to guess which property was used to create the arrangement.

What You Need

For entire class:
- ❑ 1 set of ten rocks and minerals in egg cartons (see pages 7 and 8)

For each group of 4 students:
- ❑ 1 set of ten rocks and minerals
- ❑ 1 pencil
- ❑ 1 blank sheet of paper
- ❑ 1 index card
- ❑ 4 magnifying lenses
- ❑ 1 tray (If trays are unavailable, large sturdy paper plates may be substituted. This is the case throughout the unit, whenever trays are listed.)

Getting Ready

Place one set of ten rocks and minerals, a pencil, a blank sheet of paper, an index card, and four magnifying lenses (or one for each student) on each tray, ready to hand out.

Scratching the Earth's Surface

1. If your students do not know the term already, explain that the upper layer of the Earth is called the *crust*. Ask:

- Does the Earth's crust look the same everywhere? What are some of the natural materials in the crust?

- If you were interested in finding out firsthand what the Earth's crust is made of, how would you go about it?

2. Explain that it is possible to dig or drill down into the crust, but it's very difficult and expensive. We can learn a great deal about what the crust is made of by just "scratching its surface," to gather samples and examine them. In this unit we will do just that.

3. Tell the students they will be working in groups of four to examine samples of the Earth's crust.

Exploring Rocks and Minerals

1. Demonstrate the use of magnifying lenses: "Start by holding the lens close to the sample. Then move your eye close to the lens, and slowly move the lens back and forth until you can see the sample clearly."

2. Tell the students that they are about to receive samples of the Earth's crust that they should handle very carefully. Mention that each egg carton contains ten numbered samples. Then, distribute trays of materials.

3. Have the students carefully remove the ten samples

Strictly speaking, some of the samples are not rocks, but minerals. In the next session the students learn the distinction between them. In this session, however, it is okay to use the terms "rocks" or "stones" loosely to refer to all of the samples, or to use the word "samples" or even "pieces of the Earth's crust" to describe them all.

and place them on the tray. Ask them to: "First look at the samples without a lens. How are they the same? How are they different?" *Allow time for free exploration and discussion.* Have students carefully examine all of the samples using a magnifying lens and their fingers, including their fingernails if they wish. Ask them to discuss how the samples are different from each other.

Classifying Rocks and Minerals

1. Instruct student groups to work together to arrange all of their samples into two or more groups/categories according to one characteristic (or property). Tell them **not** to let other student groups know which sorting characteristic they have selected. Give them a few minutes to work together to arrange their samples.

2. Have each group write the property or characteristic they chose on an index card and turn that card upside down next to their arrangement of samples.

3. Have teams "tour" each others' ways of grouping the samples. When each member of the student group has had a chance to guess how the samples were sorted, they can check their guesses by reading the index card. Remind groups to talk quietly so other groups can have a chance to guess too.

4. After all student groups have had a chance to tour all the other sample arrangements, have students return to their seats. Encourage a brief discussion of the various ways to sort or classify that all the groups chose and any other issues that arose.

Introducing Properties

1. If your class is not familiar with the term, define **property** as a characteristic we can observe that allows us to distinguish one sample from another. For example, color is a property. Ask the students to name others (such as hardness, texture, shape, or size). On the chalkboard, list all of the properties that the students can think of.

Among the physical properties observable without a microscope that can be important in rock and mineral classification are: nature and degree of luster; characteristic colors left when a streak test on unglazed tile is done; degree of hardness; consistency in how they fracture when they break; cleavage, or the tendency of minerals to break in a certain direction along a smooth plane; and properties related to "tenacity" (brittleness, elasticity, malleability, etc.). Some rocks and minerals also exhibit, and can be distinguished by, unusual physical properties, such as magnetism, fluorescence, ability to conduct heat, radioactivity, and other attributes. Several of the scientific supply catalogs listed in the "Resources" section contain information and equipment related to these more specialized tests.

2. If the students have listed size or weight, ask them, "Would size be a good property for distinguishing one *kind* of stone from another?" [Probably not, because a sample could be large or small, and still be the same *kind* of stone.] "Would color be a good property?" [Yes, if a certain kind of stone always has nearly the same color.] "How about weight?" [No, because a big stone would weigh more than a small stone, but they could both be of the same kind.] Cross size and weight off the list, plus any others that do not seem to be helpful in distinguishing one *kind* of stone from another.

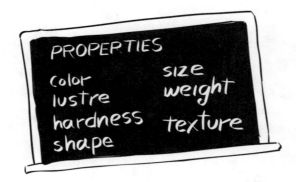

3. Inform the students that they, as scientists interested in determining what the crust of the Earth is made of, have taken an important first step. They have used their own observations and descriptions of properties of materials collected from the crust to arrange, or **classify**, specific samples into groups. Classifying the materials that are part of it is one good way to start figuring out more about the Earth's crust.

4. Ask the students if they know the names of any of the samples, and allow time for some of the students to share their ideas. Point out the "Rock and Minerals Key" attached to the inside lid of the egg carton containing their rock and mineral collections. Mention that in the next few sessions they can refer to the samples by their numbers or names.

5. Ask the students to place all of the samples back into the cartons and collect these and the magnifying lenses for use in the next session.

6. Depending on time and student interest, you may want to add a few larger ideas about the Earth's crust. You may want to draw a circle on the chalkboard to represent the Earth and roughly illustrate:

- The Earth's crust is composed of a variety of interesting "landforms," such as volcanoes, oceans, valleys, lakes, mountains, and rivers, to name just a few. All of the landforms we see are made of rocks and minerals.

- Studies worldwide of these and other landforms show that the Earth's crust is generally arranged in horizontal layers many, many miles thick, with the youngest layers, found closest to the surface, resting on top of the older layers.

- The Earth's crust has been constantly changing since its formation, over 4.5 billion years ago. In later sessions, students investigate how some of these changes may have formed and shaped their samples of the Earth's crust.

7. Invite students to bring in rocks that they find around their homes, in their neighborhoods, or near the school, so that there will be some "mystery rocks" for later in the unit.

Rocks and Minerals Key

Rocks and Minerals Key

1. schist   6. galena
2. halite   7. slate
3. granite   8. obsidian
4. quartz   9. shale
5. basalt   10. conglomerate

Rocks and Minerals Key

1. schist   6. galena
2. halite   7. slate
3. granite   8. obsidian
4. quartz   9. shale
5. basalt   10. conglomerate

Rocks and Minerals Key

1. schist   6. galena
2. halite   7. slate
3. granite   8. obsidian
4. quartz   9. shale
5. basalt   10. conglomerate

Rocks and Minerals Key

1. schist   6. galena
2. halite   7. slate
3. granite   8. obsidian
4. quartz   9. shale
5. basalt   10. conglomerate

Rocks and Minerals Key

1. schist   6. galena
2. halite   7. slate
3. granite   8. obsidian
4. quartz   9. shale
5. basalt   10. conglomerate

Rocks and Minerals Key

1. schist   6. galena
2. halite   7. slate
3. granite   8. obsidian
4. quartz   9. shale
5. basalt   10. conglomerate

LHS GEMS: *Stories in Stone*

XV

One must scour the whole coastland
of Lake Tragosoldo in Antiñana
when the hard leaf of the cinnamon
shows the first flash in the dew
and pick the drenched stones, the beach-grapes of jasper
and fiery cobble,
wet pebbles, honeycombs in
the rock, pitted
by volcano and storm
and the tusk of the wind…

Yes, the oblong chrysolite,
Ethiopian basalt,
the Cyclop's map
in the granite,
wait for you there, though nobody knows
but the anonymous fisherman, pierced
by the shuddering catch of his calling.

Only I take note of such things
in the casual morning,
I keep my appointment with stones,
slippery, crystalline, ashen,
submerged,
with hands full
of dead conflagrations,
mysterious edifices,
translucent almonds.
I come back to my family
and the day's obligations,
knowing less than I did at my birth
and grown simpler each day
with each stone.

> —Pablo Neruda
> from Las piedras del cielo (Skystones)

Session 2: Distinguishing Rocks from Minerals

In this session, students continue their investigation and classification of the Earth's crust, as they learn the difference between rocks and minerals.

The students begin by comparing two samples: one that appears to be made of a single substance and one that is clearly composed of several different substances. They then classify the other samples according to whether they seem to be made one substance or several. Through their observations, their attention is called to the essential distinction between rocks and minerals. *Minerals* are composed of a single substance, while *rocks* are composed of numerous distinct grains of different substances (such as the crystals of different minerals in the rock called granite).

Students then gain hands-on experience with a mineral, as they work in pairs to begin the process of growing halite (salt) crystals, using plastic spoons as evaporation dishes.

Note: At the end of this session, we have included an optional "Minerals at Home" handout to encourage students to think about how important rocks and minerals are in our daily lives. We also suggest one optional activity that uses this handout. Depending on your time constraints and sense of student interest, you may want to introduce the "Minerals at Home" information at a later point in the unit or devise your own ways to help students discover how much our everyday existence relates to rocks and minerals.

What You Need

For the entire class:
- ❏ 1 set of ten rocks and minerals
- ❏ 3–5 cups of near-boiling water
- ❏ 1 roll of masking tape
- ❏ 1 cup of kosher salt (Kosher salt gives best results because it dissolves without clouding, but table salt may be substituted.)

X

I invite you to topaz,
to the yellow
hive in the stone,
the bees,
and the lump of honey
in the topaz...

Pablo Neruda
Las piedras del cielo
(Stones of the Sky)

© 1974 by Grove Press Inc., and reprinted with permission. See "Acknowledgments" and "Literature Connections" for full reference listing.

For each group of 4 students:
- ❏ 1 set of ten rocks and minerals
- ❏ 4 magnifying lenses
- ❏ 2 small paper or plastic cups (about 2 oz. capacity)
- ❏ 16 stir sticks
- ❏ 1 strip of masking tape, 6" long
- ❏ 1 sturdy paper plate
- ❏ 1 tray
- ❏ 2 plastic spoons (*Note:* Spoons should be clear or black for crystals to be seen easily. If only light-colored spoons are available, put a piece of black plastic tape on the inside of the bowl of each spoon. If clear spoons are used, they work best on a dark surface, such as a dark desktop or piece of dark paper. Also see the note below on the possibility of using plastic cups instead of spoons to grow the crystals.)
- ❏ (*optional*) 2 clear plastic 9–10 oz. cups (for a larger evaporation experiment, as described at the end of this session)

For each student:
- ❏ (*optional*) 1 information sheet "Minerals at Home" (master on page 31). See #3 in "Going Further" for an activity suggestion.

Getting Ready

1. Put approximately 2 teaspoons (2/3 of a table-spoon) of kosher salt in each small cup.

2. Have 3–5 cups of very hot water available in the classroom using a hot plate or coffee maker. If a source is not available, pour boiling water into a thermos for use in the classroom.

3. For the first part of the class, "Distinguishing Rocks from Minerals," prepare to distribute to each team: a set of ten rocks and minerals, the keys, and four magnifying lenses on a tray.

4. For the second part of the class, "Growing Salt Crystals," assemble on a sturdy paper plate the following materials for each team of four students: two small cups with salt, two plastic spoons, and one pencil.

Distinguishing Rocks from Minerals

1. Ask the class, "What do you think is the difference between a rock and a mineral?" Accept their various ideas and explain that this question is what they will be exploring today.

2. Distribute the class sets of rock and mineral samples, keys, and magnifying lenses, on the trays. Hold up a sample of **halite (#2)** and **granite (#3)**. Ask the students to locate these samples and to closely examine both, using magnifying lenses. Suggest that two or three students can use magnifying lenses to inspect a sample while one student holds it at eye level.

3. Ask for volunteers to describe the samples. Accept several responses. [Emphasize that observations through magnifying lenses show granite to be composed of a number of different pieces, or *grains*, with different shapes and colors, while halite appears to be made of just one substance.]

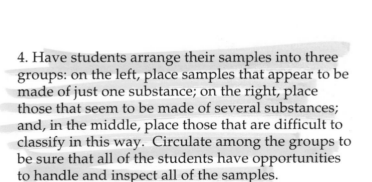

4. Have students arrange their samples into three groups: on the left, place samples that appear to be made of just one substance; on the right, place those that seem to be made of several substances; and, in the middle, place those that are difficult to classify in this way. Circulate among the groups to be sure that all of the students have opportunities to handle and inspect all of the samples.

For older students you may wish to define terms:

Organic *substances are composed of matter that is alive or was once alive.* ***Inorganic*** *substances were never alive.*

*A **mineral** is a naturally occurring inorganic substance with a certain chemical composition and set of physical properties. Many minerals occur in characteristic crystal shapes. Any material which occurs naturally but does not come from animals or plants is a mineral.*

*A **rock** is a naturally occurring solid made of one or more minerals. Most rocks contain more than one mineral, but some, such as quartzite (pure quartz) or marble (pure calcite) contain only one mineral.*

*A **crystal** is a solid substance with a regular shape. It has symmetrical plane faces, which are always at the same angle for similar sides in all crystals of the same substance. The specific crystal shape is a property of the crystalline substance. See "Behind the Scenes" for more on crystals.*

*The smallest piece of mineral in a rock is called a **grain**. Sometimes grains are in the shape of crystals, and sometimes they are not.*

5. Ask one group to report on how they classified the samples, listing their results (by number) on the chalkboard (with samples that appear to be made of one substance on the left, and samples that seem to be made of several substances on the right). Ask if other groups did it differently. Circle the numbers of samples that everyone agreed on.

6. Inform the students that samples made of just one substance are called ***minerals***, while samples made of more than one substance are called ***rocks***. Write "Minerals" above the left group on the board (made of one substance) and "Rocks" above the group on the right (made of several different substances).

7. Tell the students that it is sometimes hard to tell if a sample is a rock or mineral because it is difficult to see individual ***mineral grains*** if they are very small. Later on they will learn more about these particular samples.

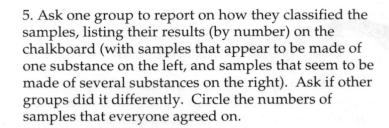

Inorganic   Organic

Mineral

Rock

Crystal

Grain

8. Explain that minerals are often in the form of regular geometric shapes called *crystals.* Ask the students to look at the group that they just classified as minerals and pick out the ones that look like they might be crystals [#2 halite, #4 quartz, and #6 galena].

9. Ask the students to place all of their samples and lenses back onto the trays. Have a student from each group put the trays with materials in a designated area.

Growing Salt Crystals

1. Remind students that the rocks and minerals they've just examined are part of the Earth's crust—so the crystals they've just looked at must have formed *inside* the Earth. Ask them for their ideas about how crystals might form. After they've shared a few ideas, tell them that in the next activity they will actually grow a cluster of crystals from a solution of water and a mineral they use every day—sodium chloride or table salt.

→ show overhead

? How do crystals form?

2. Demonstrate the procedure students will use to begin to grow the crystals, clearly explaining each step. Pairs of students should work together within the teams of four, as follows:

   a. Each pair should write both of their names on a piece of masking tape that will be used to fasten down the spoon. They should then tape the handle of their spoon to the rim of the paper plate so the bowl is flat. (This forms an **evaporation dish**—a tiny dish that allows water to evaporate within a day or two). Two spoons should be taped to each plate.

   b. Very hot water is poured into the cup until it's about half full.

c. Stir the solution in the cup for about a minute, until the salt dissolves.

d. Pour some of the solution into the bowl of the spoon until it's about two-thirds full.

e. **Keep the excess solution** in the cup and put it on the plate, next to the spoon.

3. When students understand what they are to do, have one student from each pair line up, ready to take their materials. Pour about one ounce of hot water into each cup containing salt, put it on the tray, and have the student take the materials back to her partner.

4. Allow time for each student pair to complete the procedure. When they are done, ask students to **carefully walk** their plates of solutions to a designated location in the classroom (a well-ventilated area) suitable for overnight drying. Tell the students that they will look at their evaporation dishes tomorrow.

Some teachers of younger students prefer to have the students leave the evaporation dishes and cups on the tables, for later removal to a designated location by the teacher.

5. Ask the students what they think will happen overnight. If the students predict that crystals will form, ask them to draw the shapes that they think the crystals will have.

6. Remind students to bring in "mystery rocks" from around their homes or neighborhoods. If students have already brought some "mystery rocks" in, take a few minutes to pass them around, asking students to consider whether they are most likely to be rocks or minerals.

Going Further

1. You can have your students experiment with different evaporation dishes to see if the crystals come out differently. Place a suitable clear plastic cup upside-down and pour the solution into the depression on "top" of the upside-down cup. (Not all cups have an indentation that is suitable.) How long do the crystals take to form? Do the shapes or sizes of the crystals differ?

2. Older students can design experiments with controlled variables to focus on the most important factors involved in how quickly this solution evaporates, and the sizes of the crystals. They can vary not only the shape of the evaporation dish, but the amount of solution used, and where it is placed in the room for drying.

3. See the "Optional Activity Suggestion" summarized on the next two pages. Depending on your own curricular emphases, an activity such as the one suggested on the importance of rocks and minerals can serve as a springboard for discussion of related issues, including major environmental and ecological concerns. For example:

This activity can be flexibly adapted to the level of your students and used with the "Minerals at Home" sheet as outlined below, or you can use the sheet in connection with a homework assignment, as general information, or in any way you choose.

- Students can discuss certain mining practices considered harmful to the environment (such as strip mining) and determine why it occurs and what the alternatives are.

- Recognizing that nearly all of the products that people use are derived from either farming or mining, what is the effect of the world's increasing population? How does it affect the demand for natural resources? What is happening to the environment on a worldwide level?

Optional Activity Suggestion:

Why Are Rocks and Minerals Important?

Gather together a collection of ordinary household objects, like the ones on the "Minerals at Home" information sheet (on page 31). One teacher brought the following in a large cardboard box: light bulbs, a roof shingle, waxed paper, a nickel, a penny, a marble, a hot plate, a plastic spray-bottle, tarragon spice in a container, cereal, fork, paper clips, a piece of asphalt, pipe fittings, a pair of glasses, wire cutters, a wooden clothespin, scotch tape, a thermometer, a pencil, a cup (pottery), and a sponge.

Ask students why they think it is important to study rocks and minerals. Accept all their responses, and list these on the board, with any ideas you may want to add. Then display the materials you brought to class, and ask: "Do you think these materials are related to rocks and minerals in any way?" Model how one can "work backwards" to the source of the components or ingredients to see whether or not it has a connection to rocks and minerals by asking questions such as: "What materials are needed to make this object?" "Where do these materials come from?" Help the students follow the trail of materials further and further back until they were mined (or grown) in the Earth's crust.

Students could help you sort the materials into three piles: (1) Objects or materials that include rocks and/or minerals. (2) Items that do NOT contain rocks and/or minerals, and (3) Not Sure. Encourage the students to debate the sort as needed, and provide information that may help them. For example: Glass is made from sand, which is weathered rock. All metals come from metal ores, which are rocks or minerals that contain metals. Plastic is a petroleum product that comes from oil wells, coal mines, or other types of mines. (*Note*: Strictly speaking, petroleum is not a mineral product because it derives from crude oil, which is the partially-decayed remains of plants and animals. This, however, is a subtle point, and is best mentioned to students already aware of the distinction between organic and inorganic materials.) **The main idea is to help students discover how many daily items are connected to rocks and minerals. Don't be concerned if there are some items that no one is sure how to classify. You might want to have students do further research on these.**

Distribute the "Minerals at Home" information sheet to each student, and give them a few minutes to look at it. You could ask students if they can find similar things made of rocks or minerals in the classroom, or at home. (You may want to mention that some of the items listed on the "Minerals at Home" sheet are not, strictly speaking, minerals. Some are chemical elements, such as metals, which are contained within minerals. Also, note that two large items, cars and televisions, have many minerals in them, but these are not detailed on the sheet. How could one find out what minerals these objects contain?) In conclusion, you may want to ask students to discuss or write about this question: "What would life be like for us if we could no longer mine rocks and minerals from the Earth?"

MINERALS AT HOME

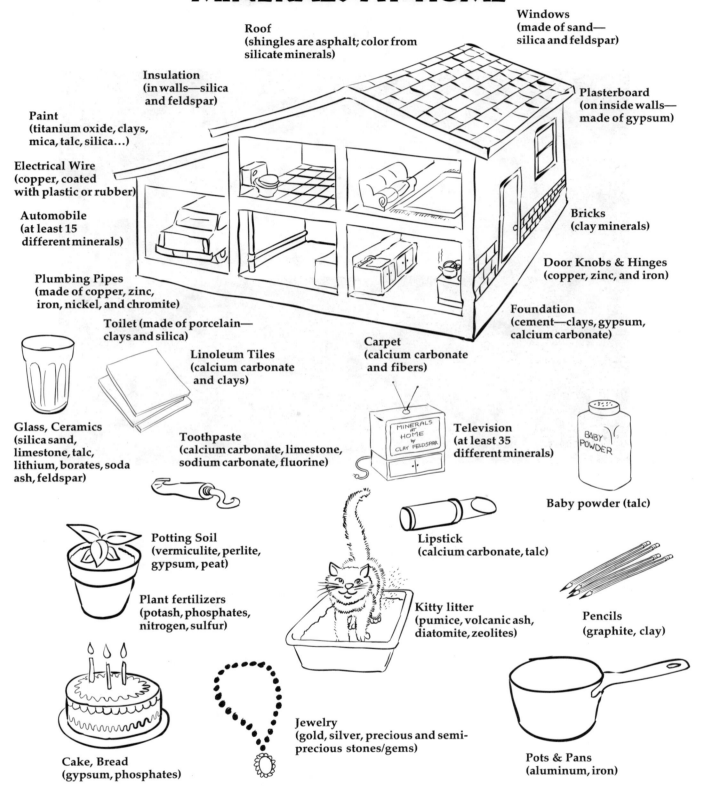

Roof
(shingles are asphalt; color from silicate minerals)

Windows
(made of sand—silica and feldspar)

Insulation
(in walls—silica and feldspar)

Plasterboard
(on inside walls—made of gypsum)

Paint
(titanium oxide, clays, mica, talc, silica…)

Electrical Wire
(copper, coated with plastic or rubber)

Automobile
(at least 15 different minerals)

Bricks
(clay minerals)

Plumbing Pipes
(made of copper, zinc, iron, nickel, and chromite)

Door Knobs & Hinges
(copper, zinc, and iron)

Toilet (made of porcelain—clays and silica)

Foundation
(cement—clays, gypsum, calcium carbonate)

Linoleum Tiles
(calcium carbonate and clays)

Carpet
(calcium carbonate and fibers)

Glass, Ceramics
(silica sand, limestone, talc, lithium, borates, soda ash, feldspar)

Toothpaste
(calcium carbonate, limestone, sodium carbonate, fluorine)

Television
(at least 35 different minerals)

Baby powder (talc)

Potting Soil
(vermiculite, perlite, gypsum, peat)

Lipstick
(calcium carbonate, talc)

Plant fertilizers
(potash, phosphates, nitrogen, sulfur)

Kitty litter
(pumice, volcanic ash, diatomite, zeolites)

Pencils
(graphite, clay)

Cake, Bread
(gypsum, phosphates)

Jewelry
(gold, silver, precious and semi-precious stones/gems)

Pots & Pans
(aluminum, iron)

AND LOTS MORE!

Session 3: The Shapes of Mineral Crystals

In this session, students gain concrete, constructive experience with some of the mineral crystal shapes found in the Earth's crust. They construct paper models of some of these shapes during class, and use the models to enhance their observations of the class samples of rocks and minerals. At the end of the session, they are challenged to use the knowledge they've gained in making the model crystal shapes to identify the crystals they've grown as either quartz or halite. This provides a direct illustration of how crystal shapes are used to help identify minerals.

Because a single crystal represents the smallest component of a mineral, careful observation and analysis of distinctive crystal shapes has proven to be one of the best ways to classify and distinguish between different minerals.

Interestingly, a limited number of crystal shapes have been found in nature. There are only seven main groups, or "crystal systems," into which all naturally-occurring crystals can be placed. This suggests that there are a limited number of ways in which atoms may be arranged together to form these shapes.

What You Need

For the entire class:
- ❏ 1 set of ten rocks and minerals
- ❏ 1 completed "Cube" and "Hexagonal Prism and Pyramid" (masters on pages 40 and 41)

For each group of 4 students:
- ❏ 1 set of ten rocks and minerals
- ❏ 4 magnifying lenses
- ❏ 2 rolls of transparent tape
- ❏ 1 tray
- ❏ crystals grown in their evaporation dishes (from Session 2)

For each student:
- ❏ 1 pair of scissors
- ❏ 1 pencil
- ❏ (optional) 1 ruler

(Copy the following onto card stock)
- ❏ 1 paper crystal model student sheet "Cube" (master on page 40)
- ❏ 1 paper crystal model student sheet "Hexagonal Prism and Pyramid" (master on page 41)
- ❏ (optional) the other four crystal model handouts (masters on pages 42–45)

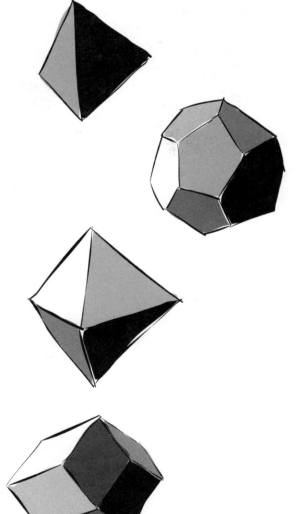

Getting Ready

1. Assembly of the crystal shapes will be much easier if you copy the sheets onto card stock instead of paper. For younger students, each shape may be copied onto a different color of card stock. For older students, the card stock is not essential, but it does make assembly easier.

2. Make one of each of the two crystal models (as described in the instructions on the sheet) to become familiar with the construction process, and so you can show the students what they will look like when completed. If you have access to an enlarging photocopier, you can make larger display crystals by photocopying the pattern in sections that can be taped together to make one large sheet. It will be easier for the students to see this larger display model.

3. Assemble all necessary supplies. Make sure that the sets of rock and mineral samples are complete. Put group supplies onto trays, ready to distribute.

 **Introducing Crystal Shapes**

1. Encourage students to examine the crystals they began growing in the previous session. Ask one or two of the students to describe what happened to their salt solution since the last session. Let students know they will be looking at these crystals in greater detail later in the session.

2. Briefly review the following definitions:

- "What's a mineral?" [a naturally-occurring substance, often with a regular crystal shape]

- "What's a rock?" [a naturally-occurring solid made of at least one mineral and usually more than one]

- "What's a crystal?" [a solid, with flat faces and a regular geometric shape]

3. To test the students' understanding, you may want to ask them: "Do minerals contain rocks, or do rocks contain minerals?" [Rocks contain minerals, but minerals do *not* contain rocks.]

4. Mention that the shapes of crystals can give us clues to help identify and classify minerals, so next the students will construct a few different crystal shapes.

Constructing Crystal Models

1. Tell the students that in a few minutes they will receive materials to construct models of two different crystal shapes, but first you're going to give them some pointers.

2. Briefly explain how to construct the cube, one step at a time, as follows:

> a. Write your name on the shape.
>
> b. Use scissors to carefully cut the cube pattern along the solid lines.
>
> c. Use the sharp edge of a desk or table to make a fold along all dashed lines of the cutout. In making the folds, they should make sure the dotted lines and any words (such as "CUBE") are on the outside of the shape being formed.
>
> d. Fit the faces of the crystal shape together, tucking under tabs as needed, and matching corresponding numbered corners. Tape the edges.

Students will have a cube when they finish. The same procedure applies to the construction of the hexagonal prism.

Older students may be able to simply follow the step-by-step directions on the student sheets.

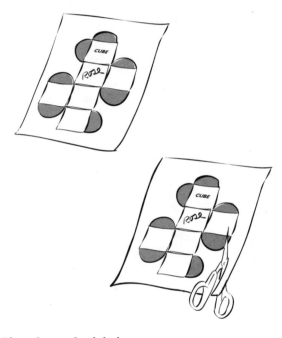

If no sharp-edged desks are available, students can use a ruler or just fingers to make the folds.

3. Tell students to go ahead and construct both shapes, starting with the cube.

4. Distribute tape, rulers, pencils, and scissors to each group. Distribute two crystal model sheets to each student: one "Cube" and one "Hexagonal Prism and Pyramid."

5. Help individuals as necessary, reminding them of steps they may have forgotten. Also remind students to write their names on the crystal shapes. As you circulate, ask: "How many faces does the cube have?" "What is the shape of each face?" "How many faces does the hexagonal prism have?"

Comparing Crystal Models with Mineral Samples

1. When all students have completed their cubes, and after most have completed the hexagonal prism, distribute trays, sets of ten rock and mineral samples, and magnifying lenses.

2. Have students place all the rock and mineral samples on the trays. Ask them to examine their paper models and compare them to the samples. Ask, "Do any of your samples have similar shapes?"

3. Ask, "Which sample or samples are most similar to the shape of a cube?" "Which are most similar to the shape of an hexagonal prism?"

4. Make sure that students identify halite and galena as having cubic shapes, and quartz as having an hexagonal (six-sided) prism shape. Explain that even though individual samples may sometimes be broken or distorted, if you look at them closely with a magnifying lens, you can almost always find that halite and galena crystals are cubic in shape, and that quartz crystals have the form of hexagonal prisms. These characteristic shapes are among the *properties* that are used to identify these minerals found in the Earth's crust.

Examining Previously Grown Mineral Crystals

1. Ask one member of each team of students to go get their crystals. Encourage the students to examine them closely, using magnifying lenses if they wish. Remind them that to produce these crystals, they used a solution of water and salt. After a few minutes of observation, invite students to describe what they have observed. [The students will find that their crystals are cubic in shape.]

2. Tell the students that different samples of the same mineral generally have the same shape. Ask them to compare the crystals they grew with the samples in their set of rocks and minerals. Is there a mineral crystal in the set which is similar? Which one? [halite]

3. Explain that **halite** is a mineral which occurs naturally in the Earth, a naturally-occurring salt crystal. Ask, "How do you think halite might be formed?" [Salt water from the oceans or some other source collected in a pool, where it evaporated, leaving the halite crystals.] Ask the students if they have any idea where salt that they use at mealtime comes from. [Some is from evaporated sea water, but most table salt is from halite that is mined, then dissolved and re-crystallized to remove impurities.]

4. Summarize by mentioning that the shapes of mineral crystals within rocks provide clues about which minerals the rock contains, and therefore what kind of rock it is. In this activity the students learned how at least one natural mineral is formed within the Earth's crust—halite. In the next session they will learn how other natural crystals are formed.

5. Remind students to continue to bring in "mystery rocks."

cube: *salt, galena, platinum*

hexagonal prism: *quartz*

tetrahedron: *chalcopyrite (a copper mineral)*

octahedron: *gold, platinum, magnetite, diamond*

dodecahedron: *gold*

pyritohedron: *pyrite (fool's gold)*

Going Further

1. Have students inspect a sample, or samples, of regular table salt. Most regular table salt exhibits cubic crystals when inspected by a magnifying lens. (Sometimes table salt crystals have rubbed together, rounding the grains.) Hand out small samples, a black sheet of paper, and a magnifying lens to your students. Ask the students if they can see the shapes of individual salt crystals, and to compare the shapes with the large crystal of halite in their set of rock samples.

2. Have the students use their magnifying lenses to study any "mystery rocks" that they may have brought in, looking for evidence of crystals.

3. Consider having students construct the four optional shapes provided on pages 42–45. If older students have constructed all of these optional shapes, then consider the following additional extension. Ask the students to count the faces, edges, and vertices (where edges meet) of the six shapes they made (listed below) and display the information in a table. Challenge students to find a relationship between the number of faces, edges, and vertices.

	Faces	Edges	Vertices
Cube			
Hexagonal Prism			
Tetrahedron			
Octahedron			
Dodecahedron			
Pyritohedron			

Note: After trying to find a relationship on their own, your students might want to know about Euler's Theorem, which can be written in two ways:

Number of faces + number of vertices = number of edges +2
Number of faces + number of vertices - number of edges = 2

XXVIII

The square in the crystal
Falls back in its symmetry:
those who open the doors of the earth
will find in the darkness, intact and complete,
the light of that system's transparency.

The salt cube, the triangular
fingers of quartz: the diamond's
linear water: the maze
in the sapphire and its gothic magnificence:
the multiplication of rectangles
in the nut of the amethyst:
all wait for us under the ground:
a whole buried geometry:
the salt's school: the decorum of fire.

—Pablo Neruda
from Las piedras del cielo (Skystones)

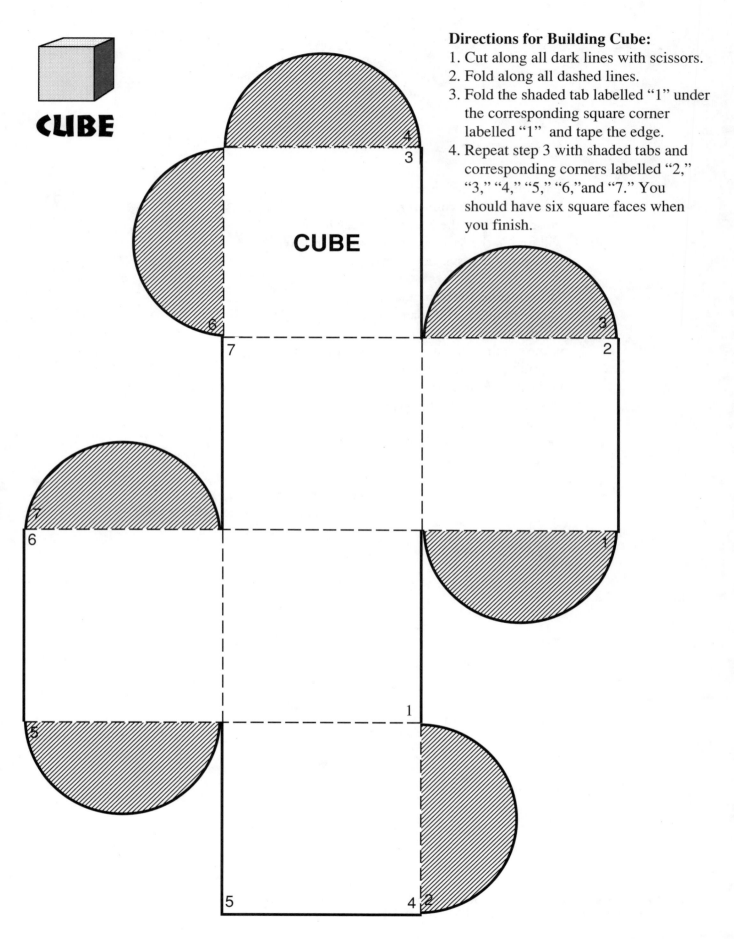

CUBE

Directions for Building Cube:
1. Cut along all dark lines with scissors.
2. Fold along all dashed lines.
3. Fold the shaded tab labelled "1" under the corresponding square corner labelled "1" and tape the edge.
4. Repeat step 3 with shaded tabs and corresponding corners labelled "2," "3," "4," "5," "6,"and "7." You should have six square faces when you finish.

CUBE

HEXAGONAL PRISM AND PYRAMID

Directions for Building Hexagonal Prism and Pyramid:

Cut along all dark lines with scissors.

Fold along all dashed lines.

Fold the shaded rectangular tab labelled "1" under the corresponding rectangle corner labelled "1" and tape the edge.

Cover shaded triangle labelled "2" with corresponding triangle "2," and tape the edge.

Cover shaded triangle labelled "3" with corresponding triangle "3," and tape the edge.

Tuck semicircular tabs under corresponding triangles and tape the edges. You should have a six-sided "tube" with six-sided pyramids on the ends.

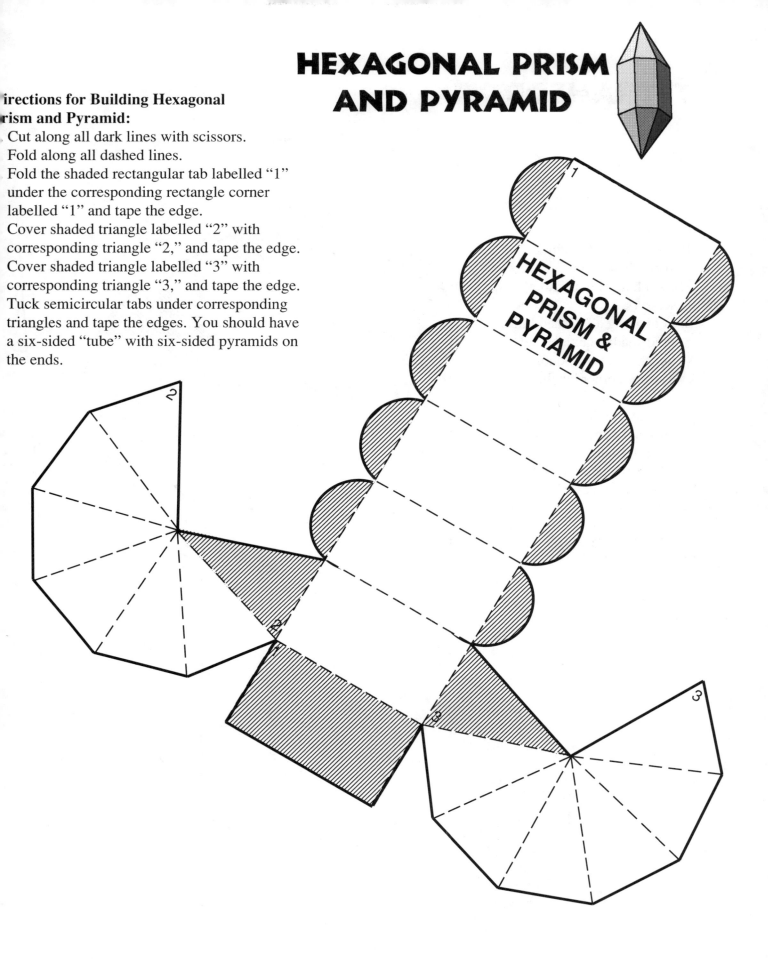

HEXAGONAL PRISM & PYRAMID

TETRAHEDRON

Directions for Building Tetrahedron:

1. Cut along all dark lines with scissors.
2. Fold along all dashed lines.
3. Fold shaded tab labelled "1" under the corresponding triangle corner labelled "1" and tape the edge.
4. Repeat step 3 with shaded tabs and corresponding triangles labelled "2," and "3." You should have four triangular faces when you finish.

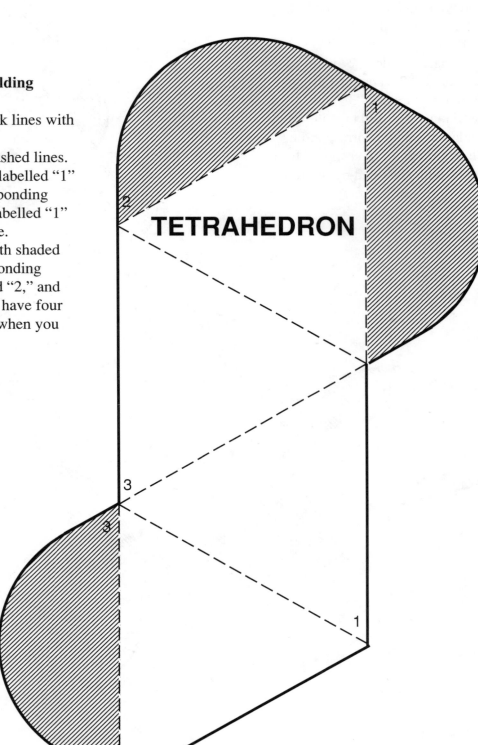

TETRAHEDRON

OCTAHEDRON

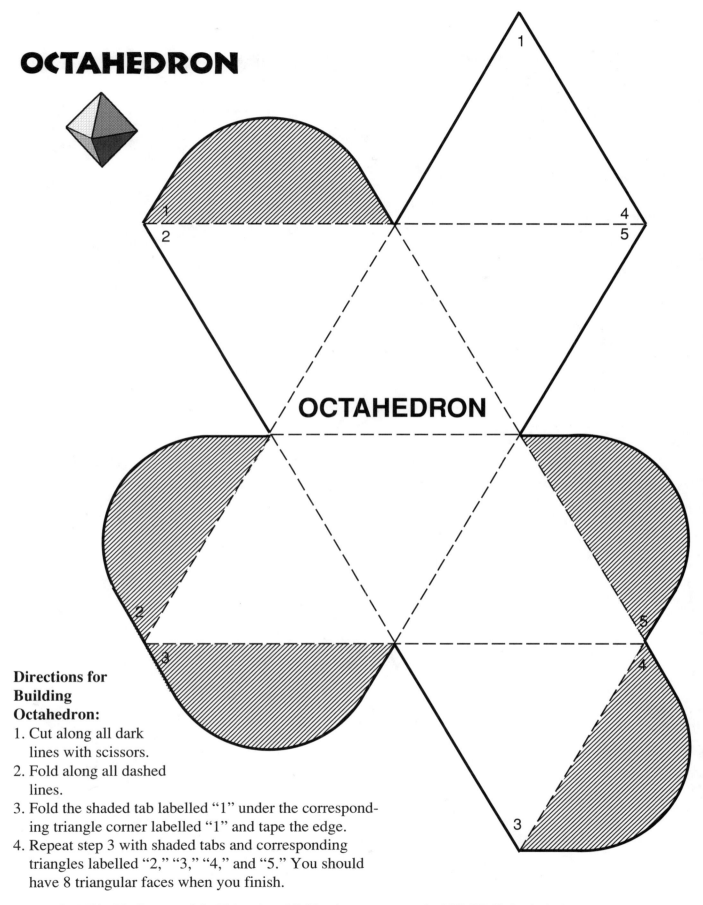

OCTAHEDRON

Directions for Building Octahedron:

1. Cut along all dark lines with scissors.
2. Fold along all dashed lines.
3. Fold the shaded tab labelled "1" under the corresponding triangle corner labelled "1" and tape the edge.
4. Repeat step 3 with shaded tabs and corresponding triangles labelled "2," "3," "4," and "5." You should have 8 triangular faces when you finish.

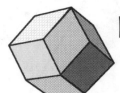

DODECAHEDRON

Directions for Building Dodecahedron:

1. Cut along all dark lines with scissors.
2. Fold along all dashed lines.
3. Put shaded tab "1" under edge "1" and tape the edge.
4. Repeat step 3 for each shaded tab and corresponding edge. You should have 12 parallelo-gram-shaped faces when you finish.

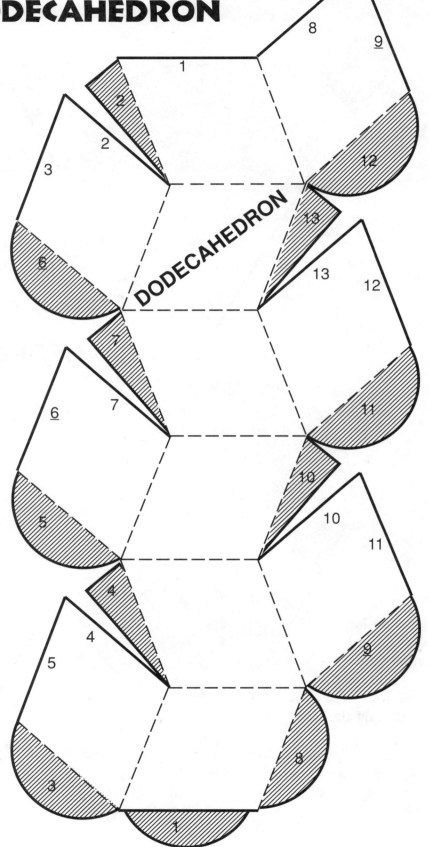

PYRITOHEDRON

Directions for Building Pyritohedron:
1. Cut along all dark lines with scissors.
2. Fold along all dashed lines.
3. Put tab "1" under edge "1" and tape the edge.
4. Repeat step 3 for the shaded tabs and corresponding edges numbered "2" through "19." You should have 12 five-sided faces when you finish.

11 12 1 13 2 14 3 15 4 16 5 17 18

4 3 5 2 1

PYRITOHEDRON

19 11 6 12 13 7 10 10 14 9 15 8 8 16 9 7 17 6 18 19

LHS GEMS: *Stories in Stone*

Session 4: Formation of Igneous Rocks

Properties of rocks, such as the characteristics of their minerals and crystals, provide important clues about how they were formed. Rocks are formed through the actions of powerful geologic processes. Three of the most important processes that shape the Earth's crust and create different kinds of rocks and minerals are: volcanism, sedimentation, and metamorphism. Each of these processes leads to the formation of a different type of rock, and rocks are accordingly classified into three major categories: *igneous, sedimentary,* and *metamorphic.* This session is about *igneous* rocks, formed through volcanism—in which molten material from the Earth's mantle rises up through the crust, where it later cools and solidifies.

In this session, to simulate the formation of igneous rocks, the students melt phenyl salicylate (salol) and observe the formation of crystals. They compare crystals formed when the salol has cooled at two different temperatures. Then students apply what they've learned to identify three igneous rock samples from their sets, inferring the relative rates at which the crystals cooled.

This session is one of the most popular in this unit. We recommend that you read it through carefully and familiarize yourself with the experiment by doing it yourself. Please see the note about classroom safety and some ideas about alternate procedures.

A Note About Candles and Classroom Safety

*This activity uses candles and some teachers have understandably been concerned about safety. Yet teachers who tested the unit have told us that, when presented in the step-by-step fashion described below, this activity not only could be conducted safely, it was the highlight of the unit for many students. It has been done safely and successfully in many classrooms and we have seen students as young as fourth grade use the candles and clearly focus on the experiments. While playing with the partly-melted candle wax is also a temptation, no accidents or injuries have been reported. Obviously, the use of a candle in the classroom necessitates care and you know best how to convey this to your students and how to structure classroom activity to ensure safety. Caution is urged regarding long hair, clothing, or other risk factors. If you feel that your students should **not** use candles on their own, we suggest two alternatives. One is for the teacher, classroom aide, or parent volunteer to sit at a table with two candles, and to have each team of four come up to do the experiment while the other students are doing something else. The other alternative, which avoids the use of flames (and doesn't require spoons), with the entire class working together, is to create a single "melting pot" from which to ladle out drops for students to examine. This alternative is described on page 59.*

It is difficult to remove salol from the metal spoons, so obtain old spoons that you can keep with the kit permanently. Any salol left on a spoon will melt during the next experiment.

It is ideal if you can acquire a spoon for every student (32 for a class of 32), plus two for the teacher. Sometimes very inexpensive metal spoons can be obtained from a flea market or secondhand store. You would then need to double the number of paper towels and ice cubes for each group.

Please be aware of the safety and related regulations in your region or state. In many places, goggles are required when there is a flame in the classroom and/ or when doing chemical experiments. In general, it is best to model good safety procedures.

What You Need

For the entire class:

- ❏ 1 set of ten rocks and minerals
- ❏ 1 book of matches
- ❏ 1 votive candle with holder (a small aluminum foil dish works fine)
- ❏ 1 container of salol crystals (2 oz. is adequate for a class)
- ❏ ice cubes, enough so every team of four students has a cube (a cooler or access to a freezer is helpful)
- ❏ 2 **metal** spoons, for permanent use in this kit
- ❏ 1 quarter-teaspoon measuring spoon

Set of 10 Rocks and Minerals

Matches

Measuring Spoon

Votive Candle

Metal Spoons

Ice Cubes

Salol

For each group of 4 students:
- ❑ 1 set of ten rocks and minerals
- ❑ 4 magnifying lenses
- ❑ 1 paper towel
- ❑ 1 tray
- ❑ 2 paper or plastic cups, 2–3 oz. size
- ❑ 2 votive candles with holder
- ❑ 2 **metal** spoons, for permanent use in this kit
- ❑ 4 pencils
- ❑ 2 lumps of clay, or another method to support the spoons with salol in them. (Pencils have been used. Some teachers recommend paper plates with raised edges. Some trays already have ridges that would serve well for this purpose.)
- ❑ 4 pairs of goggles
- ❑ 4 "Observing Crystal Formation" data sheets (page 62 or 63)
  Note: A data sheet for younger students, with less text, is included on page 63, in case you prefer it.
- ❑ (*optional*) 1 flashlight

Set of 10 Rocks and Minerals

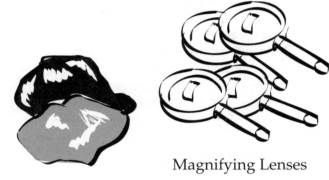

Magnifying Lenses

Lumps of Clay

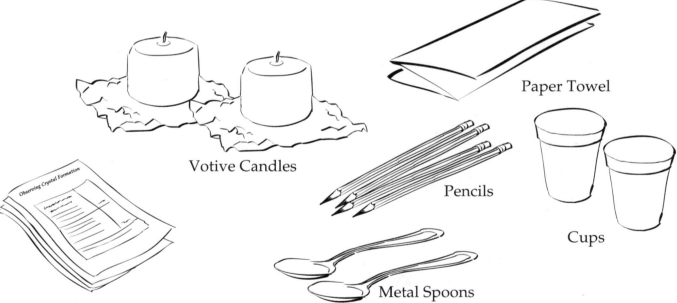

Votive Candles

Paper Towel

Pencils

Cups

Metal Spoons

"Observing Crystal Formation" data sheets

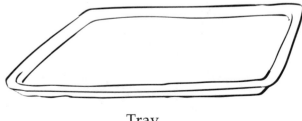

Tray

Goggles

Getting Ready

Before the day of the activity:

1. Salol (phenyl salicylate) can be purchased from science supply companies such as those listed on page 7. Alternatively, you may be able to obtain some from a high school lab. For more information on salol, please see the "Behind the Scenes" section, page 124.

2. To become familiar with the experimental procedure, carry out the following steps:

> a. Prepare the materials you will need. Light a votive candle. Set two metal spoons, a lump of clay, and a magnifying lens in front of you.

> b. Place no more than 1/8 teaspoon of salol crystals on a metal spoon. Even 1/16 teaspoon is plenty. The reason to use a very small amount is to decrease the time it takes the mass of melted salol to cool down to the temperature at which crystals start forming.

c. Heat the salol crystals by holding the spoon at least 3 cm (more than an inch) above a votive candle flame. When *almost all* crystals have melted, forming a clear liquid, remove the spoon from the flame. It is best to remove the spoon from near the flame a little **before** the last crystals melt. Enough heat will remain in the spoon to melt the remaining crystals. (If the melted salol gets too hot, it will take much longer before it cools down and starts forming crystals.)

d. Add a pinch of salol grains to act as "seed crystals." These will help start the crystalliza-tion process.

e. Place the spoon with the melted salol on a table, and position a lump of clay or pencil beneath the end of the spoon's handle, so it does not spill (or rest it on a paper plate or tray with a high ridge). Observe the melted salol with a magnifying lens as it cools.

f. Repeat the entire process a second time using a different spoon. This time, hold the spoon containing the melted salol on top of an ice cube on a paper towel, and carefully observe crystallization using a magnifying lens.

g. When near-total crystallization has occurred, place the ice-cooled spoon on the table next to the one containing crystals that are forming at room temperature, using the same support for the handles of both spoons. Compare both sets of crystals with the naked eye and with a magnifying lens.

Sometimes a large, often fan-shaped cluster of very tiny, whitish crystals forms when the salol cools very quickly. Occasionally, students see the larger shape and jump to the mistaken conclusion tha itt represents one large crystal.

3. You will probably notice that the salol placed on the ice cube cooled faster and formed smaller crystals than the room temperature salol. If you have time, remelt the salol in the spoons and try the experiment again to become more familiar with the variables that affect the outcome.

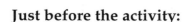

Just before the activity:

1. Place the materials you will use to demonstrate the experiment in a place where the students can easily see you.

2. Put aside the ice and sets of rocks and minerals for use later in the period.

3. Place all other supplies for teams of four students (working in pairs) on trays. For each small cup, measure out a level quarter teaspoon of salol crystals.

Good light is helpful for observations. Some teachers provide flashlights to help illuminate the crystals. If you can locate students near windows for good natural light that is ideal.

Introducing Igneous Rocks

1. Remind students how they observed properties when they began classifying the rock and mineral samples and how they've also constructed models of different crystal shapes. Explain to the students that another way to classify rocks is to study the minerals within them and their other properties to determine **how they were formed.** One kind of rock is formed when a batch of hot, liquid and crystal mush, called *magma*, cools and solidifies. Ask, "Who knows what landforms of the Earth's crust produce magma?" [volcanoes] "What do we call magma that actually reaches the Earth's surface?" [lava]

2. Explain that when magma cools it forms *igneous rocks.* Igneous rocks are one of three major classes or types of rock found in the Earth's crust. (The other two classes are *sedimentary* and *metamorphic*, which they'll have the chance to find out more about in later sessions.)

Some teachers may want to explain the main points in Steps 1 and 2 above, but then, instead of or in addition to Step 3, introduce the activities in this session with a brief story such as the one below.

3. Emphasize that some igneous rocks form **when magma cools slowly** inside the Earth. Other igneous rocks form when hot lava comes out of the Earth and **cools very quickly**. The challenge for today is for the students to work in pairs to create their own batches of hot, liquid, and crystal mush to investigate the effect that fast and slow cooling has on the formation of crystals.

Deep within the Earth, batches of molten magma stir. When a volcano erupts, some of the magma reaches the Earth's surface, on land or sea. When this lava cools it forms igneous rocks.

Meanwhile, still inside the Earth, other magma also cools, but it cools more slowly, because it is warmer inside the Earth. This magma that cools more slowly also forms igneous rocks.

Let's suppose that the molten substance we're going to work with in this activity is magma, and let's see for ourselves what might happen when it cools at different temperatures.

Observing Crystal Formation at Room Temperature

1. Tell the students that in this first part of the activity, each team of four students will work in pairs. Each pair of students will use one metal spoon to grow salol crystals and observe as they form at room temperature. Demonstrate the procedure, one step at a time, following the directions on the student data sheet, "Observing Crystal Formation," on page 62. **In your demonstration, do NOT actually light the candle or melt the salol, but go through the motions of each step.**

2. Caution the students to do their experiments over a tray so that if hot salol spills it will go on to the tray. Tell the students that each pair of students should pour *less than half* of the salol crystals from the cup into their spoon. Tell them that they will need leftover crystals in their cups to use as "seed crystals." Remind them how you added a few grains as "seed crystals" in your demonstration.

3. Show the data sheets and explain that you want **each** student to complete his or her own data sheet during the experiment.

Strictly speaking, salol is not a mineral. It is an organic compound. However, its ability to form crystals quickly as it solidifies from a melt makes it an ideal material for students to use in learning about crystal growth. See "Behind the Scenes," page 124, for more on salol.

If you were successful in obtaining 32 spoons (one for each student) they should pour less than one-quarter of the salol from the cup into each spoon.

4. Distribute the trays with materials. When both partners are ready, light the candles for them. Remind them to hold the spoon well above the candle flame. Tell the students that they should try to get as much light as possible on their spoons for best viewing. *Optional*: Provide flashlights for students to illuminate the crystals for better viewing.

5. Circulate around the class, making sure the experiments are proceeding and encouraging close observation. Allow time for the students to draw the crystals on the data sheet.

6. As you circulate ask questions to focus their observations, such as: "Does the salol seem to be forming one big crystal, or several smaller crystals?" "Do the crystals seem to have sharp edges or smooth ones?" "How would you describe the shape of each crystal?" (It is nearly impossible to get a large single crystal to grow from multiple seed crystals. So, if the students think that they have observed one big crystal, ask them to re-examine it carefully and look for flat faces and sharp edges dividing smaller crystals.)

7. Ask the students to describe how the crystals grew. "Did the crystals form all at once or a few at a time?" "Were you able to see the faces and edges move as the crystal grew?"

8. Explain that in the next part of the experiment, one of the spoons will be left with the crystals that formed at room temperature, while the salol in the other spoon will be remelted. Each group of **four** will now work together to find out what happens when the melted salol is cooled at a lower temperature.

Comparing Crystal Formation at Different Temperatures

1. Tell the students that now they will see what happens when the hot liquid and crystal mush cool more quickly in a **cold** environment.

2. Give one ice cube to each team of four, and have them place it on the paper towel. Tell them to remelt the crystals in one of the spoons, leaving the other spoon with the crystals formed at room temperature.

3. Again, the spoon to be remelted should be held above the candle until almost all the salol melts. Then the bowl of the spoon should be held so that it touches the ice cube. Remind students to add a few "seed crystals" to the spoon as it cools.

4. Encourage close observation through questioning as you circulate among the groups. Are the crystals forming? Do they seem to be forming more quickly than before? Each student should draw the crystals that are forming in the cold environment on the data sheet. Remind students to look very closely at the crystals so they do not misinterpret large multi-crystal clusters as one big crystal. After drawing the results on the data sheet, students are asked to briefly describe in writing the differences they've observed between the crystals that formed at different temperatures.

5. If you have time, the students can repeat the experiment and *time* how long it takes for crystals to form at room temperature and in a cold environment. Doing this or other experiments can help all students have a chance to do all the tasks involved, and can confirm that their results are repeatable. If students disagree about their results, or conclusions from group to group vary widely, be sure to ask students for their ideas on this in the discussion that follows.

6. When groups have finished their experiments, have them blow out all the candles. Collect the candles and other materials.

Observing Igneous Rocks

1. Ask the students to imagine that the melted substance in their spoons was volcanic magma. Ask, "Which of your magma batches completely crystallized faster—with ice or without ice?" "Which of the batches produced larger crystals?" "What other differences did you observe in the two experiments?" Help them articulate that **larger** crystals formed with **slower** cooling.

2. Ask the students to consider how their findings might apply to igneous rocks in the Earth's crust. Say, "Imagine mineral crystals in igneous rocks that formed from magma. Suppose some cooled slowly and some cooled quickly. Which ones do you think would have the biggest crystals?" [The ones that cooled more slowly.]

Crystals in igneous rocks tend to have angular shapes, more so than crystals in sedimentary or metamorphic rocks. This can be one clue in identification. While all crystals are angular when they form, those in sedimentary or metamorphic rocks have usually been subjected to other forces that tend to blunt and distort the edges of the crystals.

3. Explain that, as modeled by the experiment they did, geologists have noticed that: magma that cools very **slowly** deep inside the Earth tends to form igneous rocks with **large crystals**; lava that erupts at the surface of the Earth, or under the ocean, cools very **quickly**, and is likely to form igneous rocks with **very tiny crystals, or even no crystals at all**. Whether cooled slowly or quickly, all of the crystals in igneous rocks tend to have **angular, sharp-edged shapes**.

4. Distribute the set of rock and mineral samples to each group. Ask the students if they can tell which of the rocks are igneous. After students have had a chance to predict, point out that granite (#3), basalt (#5), and obsidian (#8) are igneous rocks. They all formed from the cooling of magma.

5. Invite the students to examine each of these rocks closely, and to put them in order, according to how fast they think the magma cooled.

6. Lead a brief discussion of their results. Inform the students that obsidian ("volcanic glass") cools so fast that crystals have no time to form!

7. Lead the students in a brief discussion of landforms on the surface of the Earth's crust where one might expect to find igneous rocks. Ask, "Where might igneous rocks be forming?" [Wherever magma reaches or comes close to the surface.]

8. You may want to challenge older students to think about the processes involved in the formation of the salt and salol crystals. You could ask, "How was the process that created salol crystals similar to the process that created salt crystals?" "How did the two processes differ?" [In both cases crystals were formed. The salt crystals formed from an **evaporated** solution of salt and water. In this case, **water** was added to *dissolve* the salt. The salol crystals formed from melted salol. In this case, **heat** was added to *melt* the salol.]

These differences have an interesting connection to rock classification that you may want to raise again AFTER your students have learned about sedimentary rocks. Rocks that contain crystals formed by evaporation are considered sedimentary, while rocks that contain crystals formed from a melt are considered igneous.

Going Further

1. The procedure in "Observing Crystal Formation at Room Temperature" says to: Add a few grains of salol to act as "seed crystals." The result is usually a mass of many crystals. Try doing the procedure again, adding only *one* "seed crystal" or *no* "seed crystals," in order to grow a single large crystal from the melt. To pick up only one "seed crystal" use tweezers, then drop the crystal into the melt. Observe the crystal growth with a magnifying lens as before.

2. Watching salol crystals grow is much more dramatic if you have access to microscopes in addition to magnifying lenses. Obtain a projecting microscope if possible, or as many individual microscopes as you can, and have students take turns watching crystal growth with the microscopes. Use glass or plastic slides and put drops of melted salol on each slide. *Tip*: To recover the salol and slides for reuse, put plastic wrap on the slides before they are used. After crystallizing, the salol easily peels off the plastic wrap.

3. Have the students use their magnifying lenses to study "mystery rocks" that they may have brought in, looking for evidence of igneous rocks, such as the presence of angular crystals packed together.

Optional Procedure for Making Salol Crystals Without Candles or Spoons

In Session 4, Formation of Igneous Rocks, groups of students use candles to melt salol in order to observe the formation of crystals at room temperature. In situations where it is strongly felt that safety factors mitigate against any student use of candles, or there are other important considerations, the following procedure is one possible alternative.

Materials and Preparation

1. Gather the same materials as listed for Session 4, except for the votive candle, metal spoons, and clay for each student group. Additionally, you will need one 6" x 6" square of plastic wrap for *each student*.

2. Prepare a melting pot for the class, using a metal or glass container. It will be nearly impossible to clean the salol out of the container, so plan on making this your permanent salol melting pot. You may use a small-size tomato sauce or tuna can, a small (4" diameter) deep aluminum pie pan, or a 400 ml glass beaker. A beaker is especially nice as students will be able to see the crystals melt.

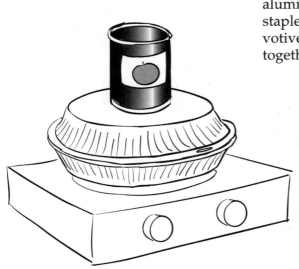

3. To melt the salol you will need a low power heat source. You could use a bunsen burner or candle with a ring stand set high above the flame, a food warming plate, or a hot plate set at its lowest setting. If you use a hot plate, make a spacer so the salol will not get too hot. You can make a spacer with two 8" aluminum pie pans placed together clam-shell style, stapled together at the edges. Or, you can make a votive candle burner from two 4" pie pans stapled together as shown in the illustration.

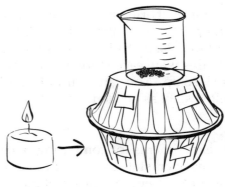

Classroom Procedure

1. Have students watch as you pour salol crystals into the melting pot and turn on or light the heat source to begin melting.

2. Distribute to each pair of students a small cup of "seed crystals" (less than 1/8 teaspoon needed), two magnifying lenses, and two squares of plastic wrap.

3. Just before the last crystals have melted, remove the melting pot from the heat source. If the pot is too hot to touch, use paper or cloth as a potholder to remove it from the heat source and wait until the pot is cool enough so you can touch it without needing the potholder. **It is important for the salol to be cool enough so that it will not melt plastic wrap.**

4. Go around with the pot and, using the quarter teaspoon measuring spoon, put a drop of melt on each square of plastic wrap.

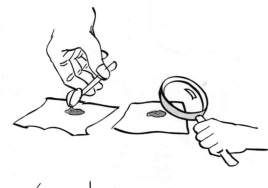

5. Have students observe the crystal growth with their magnifying lenses.

6. Students can bring back their salol crystals to the melting pot for re-melting. The solid crystals easily peel off the plastic and can be put back in the melting pot. In this way, each student can watch the melting occur close up. You can distribute additional drops of melted salol as needed.

7. The experiments for comparing rates of cooling with sizes of crystals can be conducted as described in Session 4, except that drops of melted salol can be put directly on the plastic wrap on the ice cube, rather than putting the salol in a spoon.

Name_____   Date_____

Observing Crystal Formation

Crystal Formation at Room Temperature

1. Place a very small amount (less than 1/8 teaspoon) of salol on a metal spoon.

2. Melt the salol by holding the spoon more than an inch (3 cm) above the flame.

3. Remove the spoon from the flame.

4. Add a few grains of salol as "seed crystals."

5. Prop up the handle so the spoon stays level.

6. Look at the crystals with a magnifying lens, and draw what you see.

Crystals at Room Temperature

Crystal Formation at Low Temperature

7. Remelt the crystals in **one** of the spoons.

8. Rest the bowl of this spoon on an ice cube.

9. Draw the shapes of the crystals that result when the salol cooled at a low temperature. Use the magnifying lens to compare the crystals at both temperatures.

Crystals at Low Temperature

10. Describe how the shapes and sizes of the crystals differ when they cooled at room temperature and at a low temperature.

LHS GEMS: *Stories in Stone*

Name_____   Date_____

Observing Crystal Formation

With a magnifying lens, observe the crystals formed at room temperature and draw what you see below:

With a magnifying lens, observe the crystals formed at a cooler temperature and draw what you see below:

Describe how the shapes and sizes of the crystals were different when they cooled at room temperature and at a lower temperature.

XVI

Here is the tree in the stone,
the demonstrable world, the hard beauty purely
contrived over hundreds of millions of years.
Agate, cornelian, the votary light,
exchanged fibres and juices
till the trunk of the giant
put off its dank putrefaction
and compounded a parallel image;
life failed
in the leaf,
and when the last vertical toppled,
the forest aflame, the fiery dust-cloud passed on,
all things were sealed in a heavenly cinder
till lava and time rendered
stone the reward of transparency.

—Pablo Neruda
from Las piedras del cielo
(Skystones)

Session 5: Formation of Sedimentary Rocks

In this activity, students explore the second of the three major rock types—sedimentary rocks. When rocks and minerals on the surface of the Earth's crust are exposed to wind and rain, heating and freezing, they naturally weaken and break apart. Taken together, this process is called *weathering*.

Materials formed as a result of weathering may later be dislodged and transported downhill, where they are deposited as *sediments* at the bottoms of rivers and streams, or near the places where rivers empty into large bodies of water, including oceans. This process, in which natural materials are transported from high to low areas, is known as *erosion*. Sedimentary rocks form when sediments are squeezed and cemented together under the weight of accumulated material.

In this session, students investigate sediments with different grain sizes—sand, silt, and clay—then create a model sedimentary rock profile by suspending a mixture of these materials in water, and slowly letting them settle out. Finally, they examine other soil, which they collect from an area near their school, to determine its composition, particularly with respect to relative proportions of sand, silt, and clay.

To see a World in a Grain of Sand...

William Blake
Auguries of Innocence

What You Need

For the entire class:
- ❏ 1 pound each of sand, silt, and clay in plastic bags or jars (See "Getting Ready" below for ways to obtain.)
- ❏ 2 sponges
- ❏ pitcher with 2–3 quarts of water
- ❏ newspaper
- ❏ 1 roll of masking tape
- ❏ 1 permanent marker or pen
- ❏ 1 large bucket for waste water and soil

For each group of 4 students:
- ❏ 3 small paper or plastic medicine cups (2–3 oz. size)
- ❏ 2 clear plastic cups, 6–9 oz. size
- ❏ 1 plastic spoon or stir stick
- ❏ 4 magnifying lenses
- ❏ 2 white paper towels
- ❏ 1 tray

Getting Ready

1. Sand, silt, and clay can either be obtained from natural settings (such as near rivers, lakes, or oceans) or they can be purchased from a wide variety of stores. **The differences between these materials is grain size, not their composition.** If you wish to collect the materials from natural settings, you can use wire sifters or strainers to separate by grain size. To obtain dry powdered clay yourself would involve determining likely locations, as well as warming and/or drying and powdering the moist clay. Alternatively, you can purchase these materials from various business establishments. Dry powdered clay, for example, is readily available at ceramic supply and art stores, and these materials may also be obtained at gardening, horticultural, aquarium, and hardware stores (see below).

> **Sand** grains can easily be seen with the naked eye. Sand is generally sold by gravel companies, construction suppliers, or hardware stores under the name "coarse sand."

> **Silt** grains are just visible with a magnifying lens. Silt is usually sold by gravel companies, construction suppliers, or hardware stores under the name "fill sand," "fine sand," or "half and half."

> **Clay** grains are so small that they cannot be distinguished with a magnifying lens. Clay is generally sold in art stores or ceramic supply houses. Purchase dry, powdered clay. (Ceramic supply houses may stock sand and silt as well, so could serve as "one-stop shopping" for these materials.)

2. Cover student work areas with newspaper (or have ready for students to do).

3. Have available pitcher(s) filled with water. Students can use the pitcher(s), or the teacher can circulate with a pitcher of water as needed.

Depending on where your school is located and your own preferences and schedule, you may prefer to do the local soil collection with students sometime before the session, at recess, or in some other convenient way.

4. Decide on a nearby location to take your students outside so they may collect soil samples.

5. Put about one ounce of sand, silt, and clay into three separate medicine cups for each group. Use masking tape and permanent marker to label them.

6. On a tray for each group put the sand, silt, and clay samples along with two clear plastic cups, a plastic spoon or stir stick, magnifying lenses, and three white paper towels. Set the trays aside, ready to distribute during the class.

Introducing Sedimentary Rocks

River Cutters is a GEMS guide about rivers and about how the process of erosion creates a wide variety of landforms. It could be effectively used either before or after Stories in Stone.

1. Remind the class that in the last session they focused on igneous rocks. Explain that in this session they will explore *sedimentary rocks*, which are a second major class of rocks found within the Earth's crust. Sedimentary rocks are made of *sediments*, which are fragments of older rock.

2. Ask the students, "What are some of the things that happen to rocks that might cause them to break down into tiny pieces, or fragments?" [Briefly discuss with students how wind, rain, heating and freezing, and any other natural processes students may think of, cause rocks to break down.]

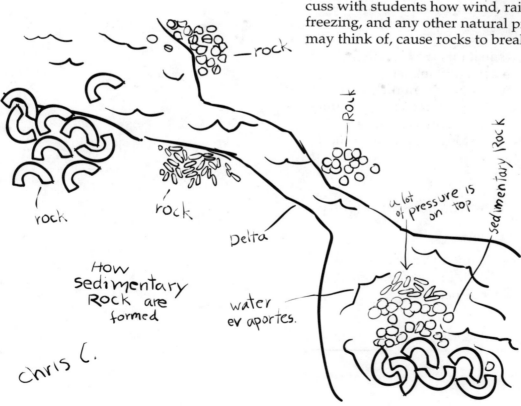

3. Explain that this process is called *weathering*. After weathering, the rock fragments, or sediments, may be transported (usually downhill) to other places by wind and rain, and by the currents in streams and rivers. The process by which sediments are stripped away by wind and rain, and transported to some other place, is called *erosion*.

Investigating Sediments

1. Tell the students that they will work in groups to examine samples of three types of sediment—sand, silt, and clay. They will first place a pinch of each type of sediment on a sheet of paper towel and look at it closely with a magnifying lens.

2. Distribute the trays of materials. Allow a few minutes for observation, including use of magnifying lenses, then lead the students in a discussion of what they have seen. "What are some of the similarities between the three samples?" "What are some of the differences?" "Are grains in the three samples the same size?" "How would you describe the *shapes* of the grains you observed?" Be sure that the students recognize the differences in grain size, and that the grains of clay are so small they cannot be seen at all. Usually, silt grains tend to be rounded. Sand grains are usually rounded too, although the shapes of the grains will vary depending on the source.

3. Inform the students that the terms *sand, silt,* and *clay* refer to grain size rather than what the materials are made of. Also mention that materials with grains larger than sand are said to be *coarse*, while those smaller than silt are referred to as *fine*.

Observing Sedimentary Layers

1. Have students slowly pour all three sediment samples into one of the plastic cups. Next, present the following scenario:

Grains of **clay** *tend to be flat, and are less than .002 mm (1/12,500th of an inch) in diameter.* **Silt** *is up to .05 mm (1/500th of an inch) in diameter, and* **sand** *ranges from 0.1 mm–2 mm (1/250th of an inch to 1/12th of an inch) in diameter. Larger pieces are called pebbles.*

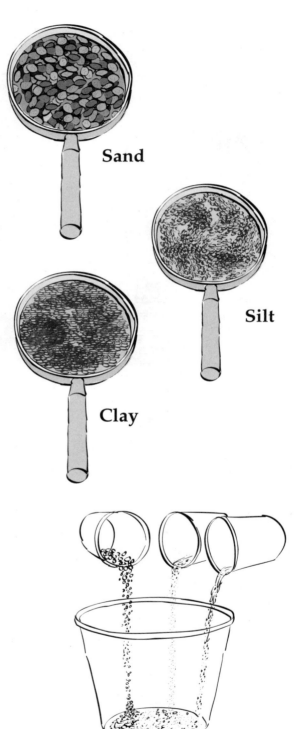

Sand

Silt

Clay

"Imagine that these sediments are fragments of rocks that once formed a tall mountain. Over millions of years, the wind and the rain wore away the once mighty mountain. A nearby river later transported these sediments many miles away from where they were first formed. At an even later point, when the river dried up, the sediments settled out of the water that transported them, and finally came to rest. That could be the story behind the sediments we're now going to examine..."

2. Tell students to slowly and carefully add water to the cups containing sediments until they are filled to within 1/2 inch (about 1 cm) of the rim. Circulate with a pitcher of water for them to use. They should then stir the cups thoroughly. Ask all students to make sure that each person in the group has a chance to stir the cup, then place the cup in the middle of the group area so it can be observed closely by all members. **The cups should not be moved again!**

3. Allow a few minutes for the students to observe the settling process, to observe their own group's "sedimentary profile." (If it is not too disruptive and there is time, you may also want them to look briefly at the results of nearby groups.)

4. Regain the attention of the entire class, and invite them to describe what they have observed. Ask them, "Which layer appears to have settled to the bottom first, the one made of coarse-grained material or the one made of finer-grained material?" "Why do you think this is the case?" Accept several responses.

5. Explain that the way the sediments are settling in their cups can be compared to the settling process that takes place when sedimentary rocks are being formed. Emphasize that one of the most important distinguishing properties of sedimentary rocks is their **layered** texture. Ask them if their observations might suggest how these layers might happen. [Sediment tends to settle out of water in layers; when it dries and becomes rock the layers can often still be seen.]

6. Point out that another important property of sedimentary rocks is that the individual grains are usually **rounded**, and are not sharp-edged crystals. Ask them why they think this might be so. [Rock grains collide with each other as they are transported as sediment in rivers and streams, so they eventually tend to become rounded.]

7. Explain that sedimentary rocks become hardened when thick piles of sediment accumulate in layers that are later buried deep within the Earth's crust. They then become compacted due to the tremendous weight of overlying layers. This compaction causes water to be squeezed out of spaces between individual grains. Certain minerals dissolved in the water are left behind to fill these spaces, and later act as a sort of cement, bonding the different loose grains together. The end result is usually layered, with rounded grains—a sedimentary rock.

8. Lead the students in a brief discussion of landforms on the surface of the Earth's crust where one might expect to find sedimentary rocks, such as the ocean floor, ancient riverbeds, lakes, canyons, and deltas.

*You may want to explain that although a sedimentary rock **usually** has rounded grains and a layered appearance, that is **not always** the case. For example, a sample taken from within a thick layer may not show layers within it, and could contain sharp bits of broken rock. Layers and rounded grains are clues—not hard and fast rules.*

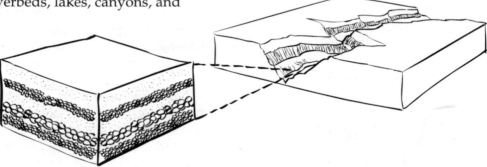

Investigating Nearby Soil

1. Ask students for their ideas about the contents of the soil near their school. Which of the three sediments (sand, silt, and clay) are in the soil? Do they think there is more or less of one kind?

2. Have each group take the unused plastic cup and plastic spoon and take them outside to where they will collect soil samples.

The GEMS unit Terrarium Habitats begins with an exploration of soil. A soil profile test using alum to separate the layers is detailed. You may want to have your students investigate soil further using such a test.

3. Outdoors, instruct the students to use spoons to scoop up two or three teaspoonfuls of soil. Suggest that different groups collect soil from different places, but all within a designated area.

4. Back in the classroom instruct students to bring their soil samples to the location where you will add water to their cups.

5. Back at their work areas, have the groups thoroughly stir their samples then put them in the center of their work area, next to the sedimentary profile that they constructed before, so that the two can easily be compared.

6. Instruct students to use their plastic spoons to scoop off material floating on top, and place it onto a sheet of paper towel. They should then look at it with a magnifying lens to see what it is made of. [This is generally organic material, such as dead leaves, twigs, etc.]

7. Next have students observe what remains in their cups. Invite students to discuss what they have observed. "Is there any layering?" "How are the layers arranged?" If they've had a chance to observe other groups, you could ask, "Is there lots of variation between these outdoor samples?"

8. Now ask the students to compare their new sedimentary profile (from the local soil samples) with the layered sediments of the sedimentary profile they created earlier. Allow a few minutes for comparisons. Ask, "What is similar and what is different about the two different samples?" "What are the relative proportions of sand, silt, and clay in soil collected near the school?" "How does that compare to the soil samples layered during the first part of class?"

9. Explain to the students that all soils consist mainly of sediments, (i.e., sand, silt, and clay) in various proportions, plus some organic material known as *humus* (partially decayed plant and animal matter). The humus is less dense than water, so it floats on top.

Going Further

1. Have your students carefully pour off the excess water and leave their soil samples for a couple of days. (It's okay—and probably unavoidable—if a little bit of matter from the top is also poured off.) Have students look at the samples again when they are completely dry. Can they see the grains? What sizes are they? What might they be made of?

2. Have students do more research and experimentation on soil composition, on what kinds of soil are considered best for growing specific plants and why, and on the interrelationships between rocks, minerals, water, and organic nutrients in the soil.

3. Your students could experiment with the best proportions for making "mudstone" to gain a sense of how sediments are cemented together in sedimentary rocks. Mix water with sand (plus clay or silt if you like) and epsom salts. Allow to dry. This could serve as a model of a sedimentary rock.

XIII

Lichen on stone: the web
of green rubber
weaves an old hieroglyphic,
unfolding the script
of the sea
on the curve of a boulder...

excerpted from a poem by
Pablo Neruda in
Las piedras del cielo (Skystones)

Session 6: Formation of Metamorphic Rocks

Metamorphic rocks, the third major type of rock found within the Earth's crust, form whenever igneous, sedimentary, or already-metamorphic rocks are subjected to intense heat and pressure.

Contact metamorphism occurs, on a generally local scale, when hot magma pushes up through the crust, transforming previously formed rocks. **Regional metamorphism** occurs, over a large area, when previously formed rocks are subjected to heat and pressure deep within the Earth's crust. In both cases, the resulting rocks exhibit properties that may be used to differentiate them from igneous and sedimentary rocks.

In this session, students create clay models which illustrate how intense heat and pressure result in properties that they can observe in metamorphic rock samples.

What You Need

For the entire class:
- ❑ newspaper
- ❑ extra set of materials for teacher demonstration

For each group of 4 students:
- ❑ 1 blank sheet of paper
- ❑ 1 plastic knife
- ❑ 1/2 bar of red modeling clay
- ❑ 1/2 bar of yellow modeling clay
- ❑ 1/2 bar of blue modeling clay
- ❑ 1 bar of green modeling clay
- ❑ 1 tray

Note: One bar of modeling clay weighs about 1/8 lb.

In this and the next session your students will be working with modeling clay. It's worth noting that this "clay" is not the clay of the last session, nor the clay that potters use that derives from the earth. Modeling clay, sometimes called modeling compound, is a plastic, synthetic product, not a natural one.

Getting Ready

1. Cover student work areas with newspaper (or have ready for students to do).

2. Assemble group supplies on trays, ready to hand out.

3. Assemble an extra set of materials for teacher demonstration. Before the day of the activity, go through all the modeling steps with the clay at least once to familiarize yourself with it.

Introducing Metamorphic Rocks

1. Inform the students that the third major class of rocks found in the Earth's crust is called *metamorphic*. Ask, "What are the other two?" [sedimentary and igneous]

You may want to mention that "meta" refers to change, and "morph" to form, so metamorphic means to change form. A closely related term, metamorphosis, is of course used in biological science to describe, for example, transformations that occur in the life cycle of many insects.

2. Explain that metamorphic rocks form when sedimentary, igneous, or already-metamorphic rocks are exposed to very intense heat or pressure. Sometimes magma pushes up through the crust, and in that case the resulting heat and pressure may change the rocks that were there before. In other cases, rocks become buried deep within the Earth, where extremely high temperatures and pressures may cause these changes to occur.

3. Tell the students that in this session they will create a clay model that demonstrates how metamorphic rocks form. You will tell them an environmental "story in stone" and, as the story is told, they work with the clay to show what happens.

Some teachers help students understand this task by mentioning "Claymation," the animation process. Each step of the clay simulations could be compared to one frame of a claymation movie, or even to a frame of a time-lapse film of geologic change.

The Formation of Metamorphic Rocks

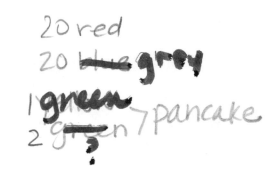

1. Explain that they should first form, with all of their red and blue clay, about 20 red and 20 blue clay spheres, about the size of marbles. Demonstrate how to do this, and how they should also form the yellow clay into a pancake shape, about the size of their hand, and break the green bar in half to make two more hand-sized "pancakes."

2. Distribute bars of different colored modeling clay, a plastic knife, and a sheet of paper to each group and have them form the 20 red and blue clay spheres, the yellow pancake, and the two green pancakes. They can use all of the red and blue clay to form the small spheres.

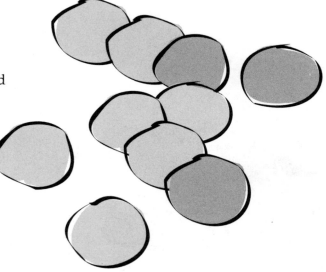

The Ancient Lake Bed

1. When all groups have finished, focus the attention of the class and have them complete each step below as you describe the Earth processes whose effects they are simulating. Read aloud or adapt the geo-logic story below, stopping to explain or repeat in your own words as appropriate. Emphasize that the story describes processes that actually occur over millions of years. Walk around the room helping as necessary, and making sure that students take turns handling the clay.

> **"Let's start by creating a model of an ancient lake bed, the bottom of a lake. That's where this story in stone begins. Into the ancient lake bed, over many thousands of years, streams carrying large amounts of greenish, fine silt have deposited their loads of silt into the lake."**

2. Explain to the students that, in the model, the green clay represents the silt. Have students place one of the green pancakes on the paper. Make sure that each group member gets a chance to flatten it a little, then continue the narrative.

blue silt

blue/white pancake

blue

"This is now an ancient layer of green-colored silt. Then, over the next few thousand years or so, rainfall increases greatly, so the amount of water flowing in nearby streams also increases dramatically. Swift moving streams and rivers are now able to wash away and transport, not only silt and sand, but even larger sediments, such as large chunks of rock, or boulders. The boulders become rounded as they roll and collide with other rocks in the stream, and they are also deposited on the lake bed, on top of finer-grained sediments."

grey

3. Have students place about 20 red or ~~blue~~ clay spheres atop the *blue* ~~green~~ clay layer. (The different colors of the spheres represent the predominance of one or another mineral.)

"Soon the river slows so it can no longer move boulders, but it still carries sand."

blue/white
green

4. Have students add a flattened layer of yellow clay to represent the deposition of sand.

"Then once again the stream flow increases during years when the rains are especially strong. More boulders are washed away, and another layer of silt is deposited."

5. Have students add more red and blue clay spheres (boulders). On top of this, have them add another green, flattened layer of clay to represent silt.

"Millions of years later, after the water in the lake has disappeared and after erosion of softer rocks surrounding the lake has also occurred, the original lake bed is exposed at the surface. A fairly young stream cuts through one corner of the lake bed, exposing the sedimentary layers that form the lake bed."

6. Have one member of each group hold both ends of a plastic knife in their hands, place it above the model, and press down quickly to cleanly cut a small portion of one side away. All students should carefully observe its layering and then put that small cut-away portion aside. **You may need to emphasize to students that, once put aside, they should leave the cut-off portion as is, since it is needed for comparison a few steps later.**

"Over the next few thousand years, a major event occurs. A large, hot body of magma slowly pushes itself toward the surface of the crust, directly beneath the lake bed. As this magma reaches higher and higher in the crust, the lake bed's sedimentary layers become heated and are subjected to tremendous pressure."

7. To simulate this pressure in their models, have **each student** in a group place the top of their fist on the model and **slowly** push down on it as hard as they can for three seconds.

"Millions of years later, state transportation officials construct a highway through the ancient lake bed, resulting in a roadcut. A few years later, another road is constructed, making an intersection."

8. Have a student in each group cut their lake bed model in half, using the plastic knife. Then have them cut it in half again. There should now be four separate sections of the model (not counting the small portion set aside earlier)—one for each person in the group. Each student should take one section, examine its metamorphosed layers, and compare it with the section that was cut before being exposed to intense pressure. **(Save these pieces for the next session!)**

Describing Metamorphic Rocks

1. Collect the pieces of "metamorphic rocks" from the students before they start to modify them further. Put them away for the next session.

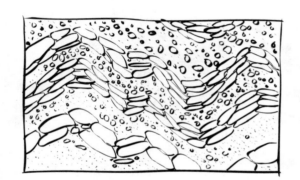

2. Stimulate discussion of what students have observed: "Can you describe the metamorphic layers?" "How do metamorphic rocks differ from igneous and sedimentary rocks?" Invite a range of responses.

3. Explain that metamorphic rocks usually have a sheet-like texture. Their mineral crystals often occur in wavy, thin bands of different colors and textures. If individual grains are visible, they generally are flat or plate-like, rather than rounded. Also, metamorphic rocks are usually much harder than the sedimentary rocks from which they are most often derived.

4. Distribute rock samples and have your students look at samples # 7 (slate) and # 9 (shale). One of these is sedimentary, and the other is metamorphic. Can they tell which is which? [Shale is sedimentary. It is made from layers of fine clay and silt. Slate is made from shale that has been metamorphosed under great heat and pressure, so it is harder than

XX

The American ranges are husky
and hairy, obdurate, snowy,
galactic
There, lives the blue mother-of-blue:
blue's secret, blue solitude,
blue's eyrie, lapis lazuli blue,
the blue spine of my country…

O cathedral of underground blue,
shock of blue crystal,
oceanic eye in the ice,
once more you rise toward the light out of water,
toward daylight,
the pure skin of space,
where blue earth is returned to blue sky!

excerpted from a poem by Pablo Neruda
in Las piedras del cielo
(Skystones)

How metamorphic Rocks are formed.

Key
sand
silt
rocks
rocks

② Ancient lake bed heavy storms, boulders wash down river layer on top of silt.

river slows, can only cary sand

③ stream flow increases more, rocks deposited on sand plus a layer of silt.

Jordan M.

④ Lake bed dries up. Thousands of years later a small stream cuts Through the lake bed.

view to cat made by stream.

- Rocks fell out
- layers "squished" Together
- sand layer "moved"

⑤ Over the next few Thousand years - hot magma pushes rowrd The earths surface. Tremendous heat and pressure!!

lake bed
crust
Heat
pressure

⑥ Millions of years pass. Make a hiway-roadcut Also make an intersection Metamorphic Rocks look like -

layer are Thinner (sheet like) minerals-be in flat plates. crystals - squished more rounded edges in thin bands harder rocks due To pressure example - shale. Slate (sedimentary) (metamorphic)

Session 7: Recycling the Earth's Crust

In this session, students are introduced to the dynamic nature of the Earth's crust by using clay to model rock cycles. They begin with the model metamorphic rock sections from an ancient lake bed which they made during the previous session. They manipulate their clay to simulate transformations of one major rock type into another, representing the effects of some of the environmental processes responsible for these transformations.

Because the Earth's crust is dynamic, always moving and changing, rocks produced by some processes are eventually exposed to numerous other processes. This means that a rock of one type may become any other rock type later on in its history. These transformations can occur over and over, and are sometimes referred to as *rock cycles*.

Underpinning the geological transformations that students model with clay in this session is the theory of *plate tectonics*, which is the cornerstone of nearly all modern thought regarding the Earth's crust. This important theory holds that the Earth's crust is broken into about a dozen rigid "plates." Large-scale geologic features such as mountain chains, fault systems, and volcanic belts are the direct result of plates of various sizes and densities moving relative to one another. As this unit suggests, **we can learn a great deal about the processes that formed these large-scale features by studying samples of rocks from which they are made**.

*The main idea in communicating the rock cycle concept to students is to deepen understanding of change and transformation in the Earth's crust. However, unlike the water cycle (as just one example) rock cycles do not **always** take place, and do not always take place in the same order. While types of rock do transform into others through geologic processes, a particular sedimentary rock may remain sedimentary for eons longer than we can imagine—it does not **automatically** change into a metamorphic or an igneous one. In fact, although there are some exceptional situations, a metamorphic rock becoming transformed into an igneous one is actually a relatively rare occurrence. **Of course, individual rocks do not change directly into another—more accurately, strata or large sections of crust are changed by geologic events, and, over time, the rocks that were or became part of that strata also reflect those processes.** While circular charts can be helpful in clarifying the rock cycle concept, and can assist students in understanding the main processes that give rise to each type of rock, they can also lead to overly mechanical or simplistic ideas that do not reflect the true natural complexity of the processes that give rise to rocks.*

What You Need

For the entire class:
- ❑ newspaper
- ❑ extra set of materials for teacher demonstration

For each group of 4 students:
- ❑ 1 blank sheet of paper
- ❑ 1 plastic knife
- ❑ 1 bar of blue modeling clay
- ❑ 1 clay model metamorphic rock section (created last session)
- ❑ 1 tray

Note: For one of the "Going Furthers," masters for transparencies of the Rock Cycle are provided, on pages 98–101. This graphic display of the rock cycle requires an overhead projector and the set of four overlay transparencies. You can make the transparencies with a copier or by tracing directly onto acetate; or, you can draw the diagram on the chalkboard or on a large sheet of butcher paper.

Getting Ready

1. Cover student work areas with newspaper (or have ready for students to do).

2. Assemble group supplies to be handed out on trays, with the clay models of metamorphic rock sections they made during the last session, along with a plastic knife, paper, and one additional bar of blue modeling clay.

3. Assemble an extra set of materials for you to demonstrate while narrating. Before the day of the activity, go through all the modeling steps with the clay at least once to familiarize yourself with it.

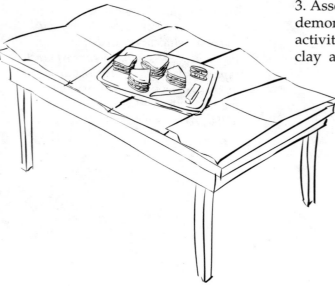

The Changing Crust

1. Explain to the students that studies of the Earth's crust have shown that it is always moving and changing. Over the past thirty years, ***the theory of plate tectonics*** has become widely accepted. It states that the Earth's crust is broken into about a dozen huge chunks, or plates, which are always in motion. Where the plates move in relation to each other—colliding, sliding past, or pushing away from each other—they create huge mountain ranges, cause earthquakes to occur, and volcanoes to form and erupt.

2. Explain that because the Earth's crust is always in motion, rocks produced by one process are often acted on and changed by other processes. In this activity, students will again work with their clay models, starting with their clay section of "metamorphic rock," to simulate the way rocks can be changed, or *transformed*, from one kind of rock into another.

Recycling the Crust

1. Distribute trays to the groups with the clay models of metamorphic rock sections they made during the last session, along with a plastic knife, and one bar of blue modeling clay.

2. Focus the students' attention on their model of metamorphic rock from the last session. Ask, "What type of rock does this represent?" [metamorphic] Say, "Beginning with this model of a metamorphic rock, here's another 'story in stone' that may help us see what might happen to rocks over time."

3. Begin the geological narrative that begins on the next page. **Take your time**, going back over and/or adjusting the language and wording as best suited for your students, **using your own storytelling style**. Pause to demonstrate the steps as appropriate, and circulate to make sure students take turns handling the clay and that all groups have completed each stage of the changing model before the story goes on to the next step.

For more on plate tectonics, see the "Behind the Scenes" section or modern geology texts. If you feel introducing this theory or other advanced vocabulary is not appropriate for your students, adjust accordingly by explaining that the Earth's crust is in motion and adapt terminology as you think best in the geological narrative that follows.

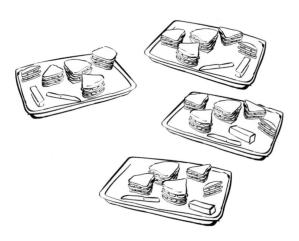

*At the end of Session 6, the ancient lake bed has a highway running through it. Some bright student may wonder what has happened to that highway, as this new story in stone begins. While it's not at all necessary to mention otherwise, in case your students raise the question, you could say that the narrative in this session could be seen as another, alternate way that the lake bed area could have developed over time and been affected by different geological events, **or** you could invite them to consider that this imagined narrative could be what might happen millions of years **in the future**—long after the highway has disappeared!*

From Metamorphic to Sedimentary

"Over the next few million years or so, the wind and rain, along with nearby streams and rivers, completely erode the ancient lake bed where these metamorphic rocks were first formed. Tons upon tons of eroded metamorphic rocks are transported to the ocean, and deposited as thick layers of sediment."

1. Ask, "What type of rock would be formed by this depositing process?" [sedimentary rock]

2. Working together in groups of four, have students carefully break off marble-sized pieces of the entire model until it is completely taken apart. Then, they should gently mash these pieces together to form five separate sedimentary layers no larger than a fist. Have them stack these layers on top of one another.

The Plates Collide

"These sediments lie on top of the place where two huge plates come together—rocks that make up the continent, or continental crust, and those that make up the ocean bottom, or oceanic crust. During the next few million years, dramatic changes occur as these two plates collide.

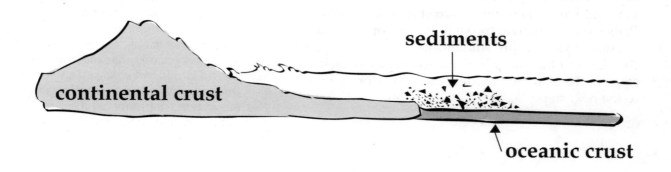

As they push into one another, the edge of the heavier oceanic crust begins to dive under the lighter continental crust. From this downward motion, the oceanic crust and the sediments which lie on top of it are pushed into deep, hotter regions of the Earth, until both are completely melted."

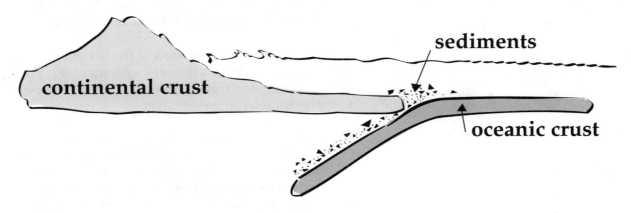

1. Ask, "What would the melted rock be called?" [magma]

2. Have students "melt," or mash, the layers together until all of the different colors mix.

A Band of Volcanoes

1. Ask the students, "How is it possible for the magma to make its way back to the surface, and where does this occur?" [It is pushed out of volcanoes as lava.] "What type of rock would this process produce?" [igneous]

2. Have students use the resulting mixed-color clay to form a model volcano, no larger than a fist.

"As a result of the collision between oceanic and continental crusts, many volcanoes have formed in a band that stretches for about 300 miles. These beautiful yet dangerous volcanoes erupt now and then over the next several hundred thousand years. Older rivers and streams, that formed before the volcanoes, are now forced to change their courses—they have to flow *around* the volcanic region to deposit sediments into the ocean."

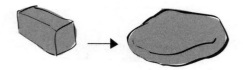

3. Have students use one third of the bar of blue clay at their tables to form a pancake no larger than a fist. They should place this pancake next to their volcano, to represent a sedimentary layer.

"Several million more years go by. Volcanic activity in this region ends when the collision between the oceanic and continental crusts lessens. At this point wind and rain begin to rapidly break down the igneous rocks that make up the volcanic region. More recent rivers and streams very slowly carve out impressive valleys and small canyons. Sediments carried by these rivers and streams are deposited into the ocean miles away, along the coast."

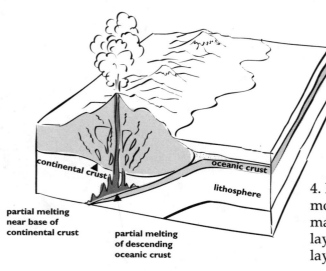

continental crust

oceanic crust

lithosphere

partial melting near base of continental crust

partial melting of descending oceanic crust

4. Have students remove about two-thirds of their model volcano, and use its clay to form 10 to 15 marble-sized spheres. Arrange these spheres into one layer by gently mashing them together. Place this layer on top of the blue pancake layer.

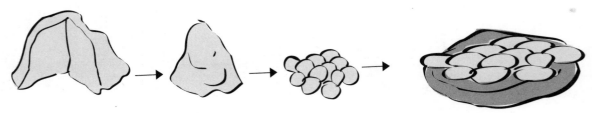

A Variety of Sediments

"Weathering and erosion of the igneous rocks from the volcanic region continue at a steady rate over the next few million years. When only about one-third of the region is left standing, older rivers carrying sediments from more distant rocks change their courses and now can flow straight through the heart of the volcanic region. Sediments transported by these older rivers differ greatly from those that came from the volcanic igneous rocks. They are more rounded and finer grained from being transported a greater distance, and they are made up of completely different minerals. At points along the coast, these older rivers deposit their sediments in layers on top of those that came from the igneous rocks of the volcanic region."

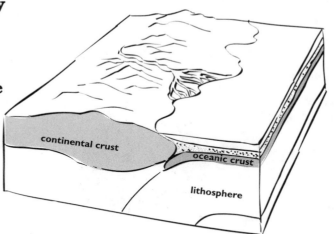

1. Students should use one-half of their remaining bar of blue clay to form a fist-sized pancake, and place it on top of the previous layer.

2. Next they use the rest of their volcano's clay to form 10 marble-sized spheres. They should carefully place the spheres on top of the last blue layer, and gently push them together from the sides to form one distinct layer.

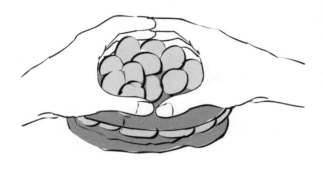

3. Have students use the rest of the blue clay to form another pancake, and place it on top of the previous layer. They should take turns gently mashing down on the new model with their fists. Caution them not to mash down too hard.

4. Next, have students use the plastic knives to cut the model into two equal parts, one for each pair of students. Encourage all students to carefully examine the layering in their sections of the model.

5. Ask the students, "If this model were actually a rock found somewhere on the surface of the Earth, what type do you think it would be, and why?" [Sedimentary. It's made of materials that have been weathered and eroded and deposited.]

6. Have students place the two sections of their model side by side on the table, and gently push them together to form one, single model.

From Sedimentary Once More
to Metamorphic

"These sedimentary layers formed along the ocean bottom are now more than five miles thick! The deepest layers have constant downward pressure on them, from the massive weight of the sediments on top of them. Eventually, these deepest layers harden to become sedimentary rocks, such as sandstone and shale.

As these materials continue being deposited, the tremendous weight of the growing sedimentary pile causes downward warping of the crust underneath. This allows the pile to hold still more materials from land. For every three feet of sediments received, the crust below sinks about two feet!"

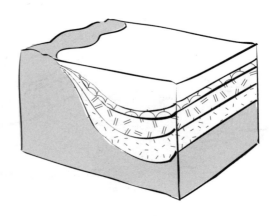

1. Have students extend their fingers out away from their wrist, place the bottom of their hands on top of the model, and press down very gently. Do this one student at a time so that every student in the group has had a turn, then stop. (This represents the downward pressure.)

"Now, collision between oceanic and continental crustal plates begins again, and the thick sedimentary rock pile that has accumulated over millions of years undergoes huge changes. Due to the powerful force exerted when the oceanic crust pushes against it from the side, the entire sedimentary rock pile is tightly squeezed together. Layers near the bottom of the pile become heated and partially melt as they are pushed down further into the crust. Because of the tremendous heat and pressure, the sedimentary rocks are severely deformed so that some of their minerals flow and accumulate in thin, wavy bands."

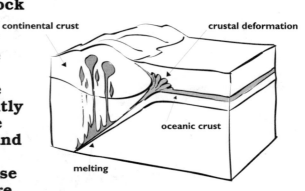

continental crust

crustal deformation

oceanic crust

melting

2. Have two students work with the model first. They should place their arms on the table on opposite sides of the model, facing each other, and make a fist. Then, at the same time, they should push both sides of the model into one another, as firmly as possible, for a count of three. Repeat this as many times as necessary for each group member to have a turn, then have students briefly examine the model.

"Throughout the next few million years, the greatly deformed rocks in the pile begin to break apart. Soon many giant cracks form. These giant cracks are called faults. Because of the faults, most of the movement within the pile is upward, instead of horizontal. Many millions of years pass. By now, so much vertical movement along faults has occurred that rocks once buried many miles below have been pushed up above the surface of the crust. These rocks eventually form the jagged peaks of a new mountain range located along the coast."

3. Have students use the plastic knife to cut the model into two equal parts. Working in pairs, they should take one section of the model and examine it closely.

4. Ask, "If this were a rock found in the jagged mountains along the coast, what type do you think it might be, and why?" [Metamorphic, because it has been subjected to heat and pressure.] (Metamorphic is the same type of rock with which this activity began.)

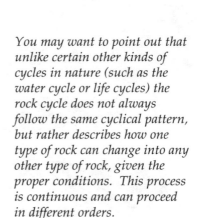

The Rock Cycle

1. Encourage a discussion about the processes described throughout this session. Be sure the students understand that these changes occur over **extremely long periods of time**.

2. As needed, clarify for them that, as they demonstrated in their modeling, a single, individual rock doesn't suddenly change from one kind to another—instead many rocks that are part of a larger layer or strata are acted on by geological and/or environmental forces, and in the process, they can be transformed into other types.

3. Point out that they began with model metamorphic rocks, which were then transformed into sedimentary rocks, then igneous, then sedimentary again, and finally back into metamorphic rocks. In this way their models have demonstrated that, as a result of the many processes shaping the Earth's crust, rocks often go through a series of transformations throughout their long histories, encountering new environmental and geological conditions, to become different rock types. Let students know that any series of transformations such as this can be thought of as being part of a *rock cycle*.

You may want to point out that unlike certain other kinds of cycles in nature (such as the water cycle or life cycles) the rock cycle does not always follow the same cyclical pattern, but rather describes how one type of rock can change into any other type of rock, given the proper conditions. This process is continuous and can proceed in different orders.

Going Further

1. **Where in the world are rocks formed**? Using a cutaway view of the Earth or a poster showing different landforms, ask students to point to places where rocks may be formed and where they might be found now. (Here are some examples from the samples in their class sets: **Conglomerate and shale** are formed in the layers of sediment from rivers and streams. It is found today in areas where old sedimentary layers have been thrust upward. **Slate** is formed in the deepest layers of sediment, where it was subjected to great heat and pressure. It may be found today in an exposed wall of metamorphic rock, where the once-flat rocks have become tilted or bent. **Granite** can be formed in one of the "fingers" of magma inside a volcano that does not reach the surface. It may be found on mountains. **Basalt** can be formed where magma came to the surface under water, where it cooled very quickly. Lots of basalt is found today on the ocean floor. **Schist** is formed deep within the earth, or near chambers of magma, where the heat and pressure melted some of its minerals. It can be found in exposed walls of metamorphic rock today.)

2. The Story of Ms. Terry Rock. Tell the students that today they will "interview" some rocks, and write their life stories. Ask them to select one of the rocks they have "met" during this class—one of the ten rock samples, or one of the "mystery rocks." Put the sample rock in front of them and begin the interview. Suggest that they listen to the rock very carefully as it tells its story. Have them start with questions about recent history, and go backwards through time: Where were you found? How did you get there? What were you before you became the rock you are now? What was it like when you took your current form? Where did you come from before that? and before that? Where were you born and what were the conditions like back then? Have students draw a picture about the story of their rock and share it with the class. They may want to illustrate an especially interesting event in its life, or to show several stages in its life, as in a cartoon.

3. Rock Cycle Diagrams. The following script, with overhead illustrations, represents a schematic summary of the transformations possible in the rock cycle. It can best be used as a way to prompt discussion of the geological conditions that give rise to such changes. As noted in Session 7, it is important that students understand that these transformations are not automatic, tend to take place over very long periods of time, and do not involve an individual rock transforming into another kind of individual rock, but instead are a result of geological and environmental processes acting upon larger landforms.

1. Ask the students, "What would have to happen to sedimentary rocks to turn them into igneous rocks?" Listen to their answers, then show the first overhead transparency of the "Rock Cycles," illustrating how sedimentary rocks would melt, become magma, and cool to become igneous rocks.

2. Ask the students, "Do you think the reverse could occur?" "Could an igneous rock become a sedimentary rock?" "How?" Listen to their answers, then place the second transparency on top of the first, illustrating how an igneous rock could become a sedimentary rock.

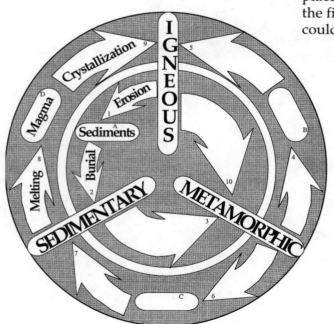

3. Do the same for the next two transparencies, asking the students to recall: (a) how igneous rock can be transformed into metamorphic rock and vice versa, and (b) how metamorphic rock can be transformed into sedimentary and vice versa. Add first transparency #3, then #4, to confirm their responses.

4. When the picture of the Rock Cycles is completed, explain that any given rock may go through one, several, or all of these different processes in its "lifetime." Also, it takes the Earth millions, and sometimes billions, of years (a very long time) to transform one type of rock into another.

The Rock Cycle

Transparency 1

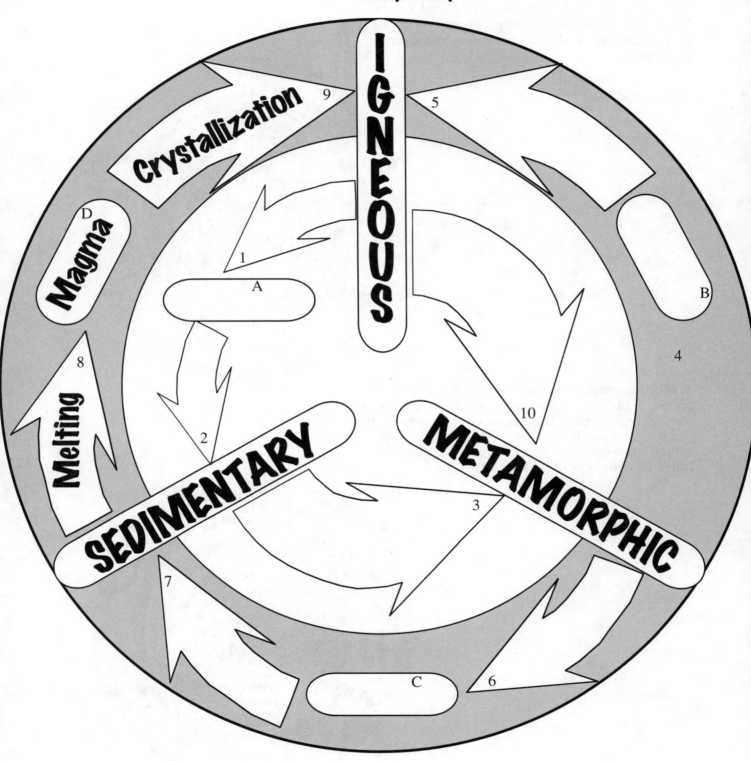

The Rock Cycle

Transparency 2

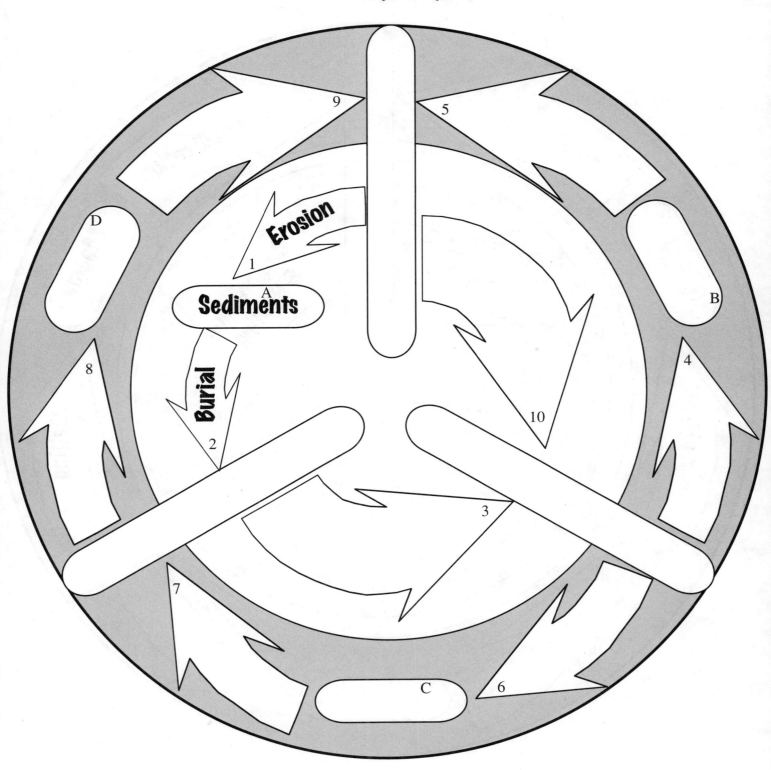

The Rock Cycle

Transparency 3

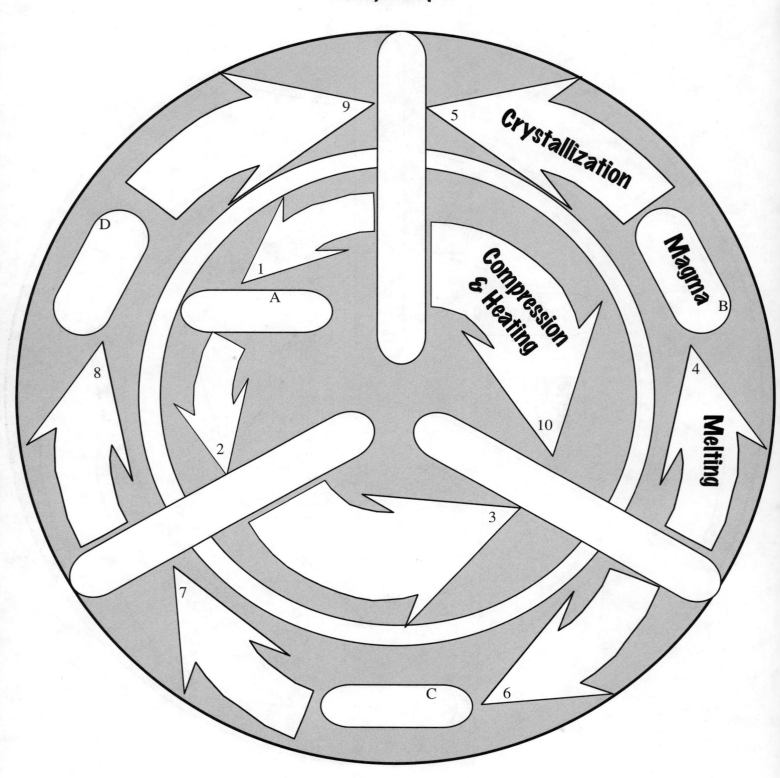

The Rock Cycle

Transparency 4

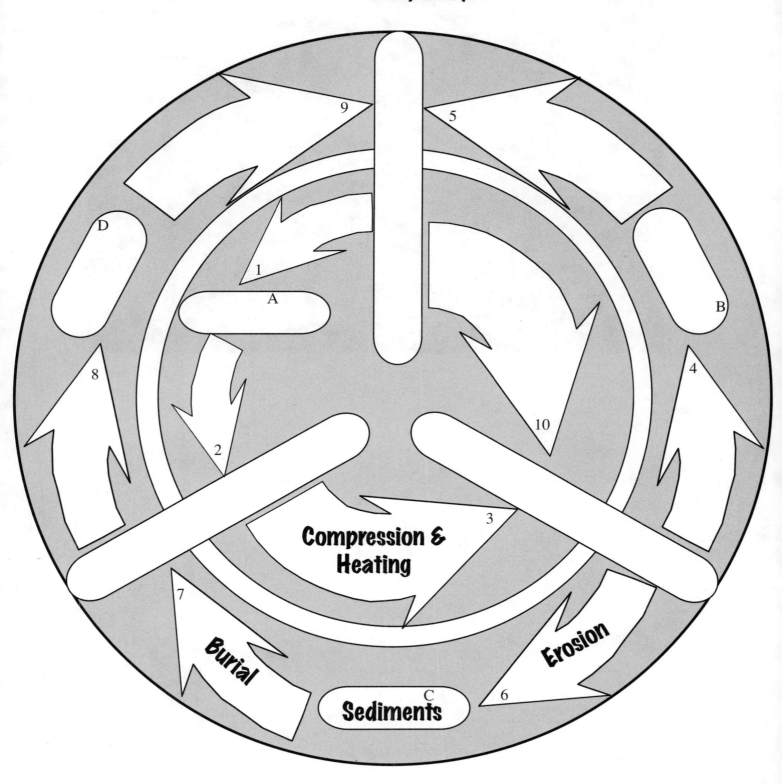

Session 8: Classifying Rocks and Minerals

In this session, students apply the knowledge they've gained to classify the ten rocks and minerals in their sets. They classify each sample as a rock or mineral, then further classify the rocks into the three basic types—igneous, sedimentary, or metamorphic. In the second part of the session, they apply their knowledge to unknown "mystery rocks" which the students have brought in—observing, recording, comparing, and inferring how these rocks may have been produced.

The objectives of this session are for your students to improve their abilities to observe and compare the properties of rocks and minerals, and to apply important concepts about how rocks and minerals form. This can happen in an educationally beneficial way whether or not students arrive at the "correct" classification of each rock. Observing students in action will give you an opportunity to assess how well they have understood the concepts in this unit.

Don't feel disappointed if your students (or you) do not succeed in identifying many of the "mystery rocks." Even trained geologists frequently encounter difficulties in identifying rocks and minerals. It is more important that students apply what they learned to make general statements about the samples, such as: "It has a crystalline shape and uniform color, so it's probably a mineral," or "It's in layers that are flat and soft, so I'd say it's a sedimentary rock." Controversy and debate about properties and attributes is not only educational—it's the stuff of science!

"Going Further" suggestions for this session include a poetic matching activity. On page 114, "Going Further" for the entire unit suggests ways for students to deepen their interest in our planet's crust. It is hoped that this unit will help spur inquisitive student minds to nurture their curiosity and continue their learning about the many facets of this one large "rock" we call the Earth.

Some classes may have already classified their samples after appropriate earlier sessions. In this case, you may want to review their classifications, and discuss other questions about classification they may have, then move more quickly into attempting to classify the "mystery rocks."

What You Need

For the entire class:
- ❑ 1 set of ten rocks and minerals

For each group of 4 students:
- ❑ 1 set of ten rocks and minerals
- ❑ 1 "Rock Type Description and Classification" data sheet (master on page 109)
- ❑ 4 magnifying lenses
- ❑ 1 "mystery rock" for each group.
- ❑ 1 tray
- ❑ 1 "Observation and Display of Mystery Rock" data sheet (master on page 110)
- ❑ 4 pencils

If you wish to supplement this unit with more information about how to identify rocks and minerals, you may want to obtain one of the field guides suggested in the "Resources" section on page 115.

Note: One of the "Going Further" suggestions features "Rock & Roll Riddles" in verse (masters on pages 111–112). Each verse corresponds to one of the samples in the class set. These poems can be used in dramatic reading and discussion, as a guessing game, a homework assignment, or in whatever ways you think best for your students. See the "Going Further" #2 on page 108.

Getting Ready

1. Place the sets of rocks and minerals with name key and magnifying lenses on trays for each group. Set them aside, ready to hand out.

2. Select one "mystery rock" for each group. You may want to choose a large sample that is different from any in the set of samples, but has characteristics that will not make it too difficult to classify given the knowledge that the students have gained so far. For example, a sample with uniform color and texture, possibly with a crystalline shape, would be easy to classify as a mineral. Other good "mystery rock" samples might have obvious layers with rounded grains (sedimentary rock), or interlocking angular crystals (igneous rocks), or very thin bands of flattened crystals (metamorphic rocks). If most of your "mystery rocks" seem difficult to classify, that is also fine, so long as you temper the students' expectations with the understanding that the challenge of trying to classify them is what is most important, and that geologists might often have difficulty with certain samples.

3. Duplicate the data sheets. For each group, make one "Rock Type Description and Classification" sheet (master on page 109) and one "Observation and Display of Mystery Rock" sheet (master on page 110).

4. On the chalkboard, list rock and mineral classifications for the ten samples. Cover the names of the rock and mineral samples with a sheet of paper taped to the board, so you can lift the sheet and reveal it later in the class.

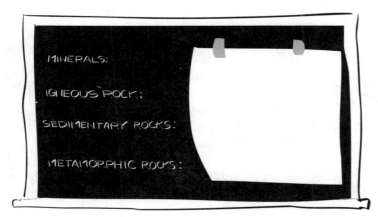

Classifying—Using What We Know

1. As a review, ask the students to respond to several questions:
- What is the Earth's crust made of?
- What is the difference between rocks and minerals?
- Can you name some crystal shapes?
- What are the major types of rocks called? How were they formed?

2. Take time as appropriate. Invite students to comment on each other's responses, and to build upon another student's ideas to mention different properties and describe possible ways rock types can form. Encourage students to raise questions they may have and ask other students to respond.

3. Distribute the sets of rocks and minerals with keys, pencils, and magnifying lenses. Also hand out one copy of the "Rock Type Description and Classification" sheet to each group.

Some teachers summarize the processes with a chalkboard diagram like this:

weathering and erosion → sediments → burial → ***Sedimentary rocks***

melting → magma → crystallizing → ***Igneous rocks***

sedimentary or igneous (or already-metamorphic) rocks + heat and pressure → ***Metamorphic rocks***

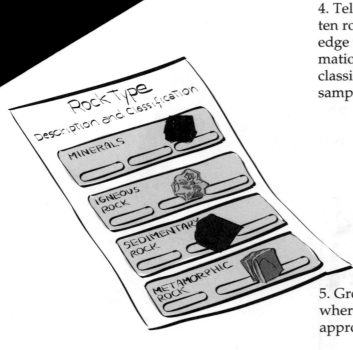

4. Tell the students to work in groups to classify all ten rock and mineral samples according to the knowledge they've gained. Point out the summary information on their data sheets that will help them classify the samples. Suggest that groups consider the samples one at a time:

> a. Pass the sample around, carefully studying it with both the unaided eye and a magnifying lens. Decide if it is mineral or a rock.

> b. If it is a mineral, put it in the appropriate spot on the data sheet.

> c. If it is a rock, they should classify it as igneous, sedimentary, or metamorphic.

5. Groups should try to come to agreement about where to place each sample, then place it in the appropriate spot on the data sheet.

6. When the students are finished, ask one group to report on the samples that they classified as minerals. Ask another group which ones they placed in the "Sedimentary Rocks" category, and so on. Briefly discuss differences of opinion.

7. Reveal the way these samples are classified geologically, by removing the sheet of paper covering the list on the chalkboard:

MINERALS: Halite, Quartz, Galena

IGNEOUS ROCKS: Basalt, Granite, Obsidian

SEDIMENTARY ROCKS: Shale, Conglomerate

METAMORPHIC ROCKS: Slate, Schist

8. Students may want to rearrange their samples as needed, and keep for comparison during the next activity.

Mystery Rocks

1. Explain that each group will now receive a "mystery rock" to classify. They should first observe the sample and record their observations—for example: Is it a uniform color? Is it composed of individual grains? Then the group will need to decide if it is a mineral or a rock. If it's a rock, they will need to decide if it is igneous, sedimentary, or metamorphic. Tell students to record their reasoning on their data sheet so that other students will be able to understand it.

2. Distribute a "mystery rock" to each group, and a data sheet, "Observation and Display of Mystery Rock." Allow about five minutes for students to attempt to classify their "mystery rock".

3. When each group has finished analyzing its "mystery rock," they should place their rock in the circle on the data sheet and leave it near the center of the table where everyone can easily see both the rock and the list of observations made by the group.

4. Ask the students to walk in pairs around the room to examine the work of the other groups. At each station they should check off one box under "Other Geologists' Opinions" to show if they agree or disagree with the group in charge of investigating that "mystery rock." Allow ten to fifteen minutes.

5. Have groups reconvene at their area and examine what additional marks were made on their data sheet. The group should then decide whether to stick with their original opinion or to change according to the opinions of others. Finally, they should choose a representative to present their findings.

6. One at a time, have the representatives hold up their "mystery rock," tell how they classified their samples and explain why they chose that classification. Those who disagree should be invited to report the observations that led them to a different conclusion. Welcome discussion and explain that classification is not always simple and straightforward. Some samples might challenge experienced geologists.

7. You may want to add that scientists often debate and discuss their findings. What students have done in attempting to classify "mystery rocks" is similar to what scientists might do at a "conference" or in graphic displays that summarize findings at a "poster session." There are often disagreements. This kind of discussion and controversy is one of the ways that science grows and changes, and that new theories emerge. As people learn more, new findings and new theories emerge. Just like the Earth's crust, science too is always changing!

Going Further

Key for Teacher:

A. Granite	#3
B. Galena	#6
C. Slate	#7
D. Basalt	#5
E. Conglomerate	#10
F. Schist	#1
G. Salt (Halite)	#2
H. Quartz	#4
I. Obsidian	#8
J. Shale	#9

1. Students could find out more about their samples, and some of the other ways they are analyzed and classified. For example, students could research the **chemical** structure of their samples, and learn more about how Earth Scientists analyze the chemical composition of rocks and minerals.

2. On pages 111–112, we include a "Going Further" matching activity that you and your students may enjoy as a "literary connection" to the processes of formation and classification of the samples. You can assign the riddles as homework, or, in class, tell groups of students to imagine that each of their ten rock and mineral samples could talk and that each of them told you a story of how it was formed and where it had been most of its "life." Their stories were written down in poetry, but you forgot to write down which stone told which story. The students' task is to figure out which sample recited which poem. If you present it in class, you may want to consider giving three or four riddles to each group, rather than all ten. Some teachers may want to do this with the entire class by selecting a different student to recite each of the riddles to the class, and having the class decide which sample is being described after each riddle is recited.

Rock Type
Description and Classification

Minerals

What It Is
Naturally occurring solids made of a single substance often found in crystal shapes.

Look for
Uniform appearance, crystal shapes, only one type of grain.

Igneous Rock

What It Is
Formed by cooling of magma. Slow cooling forms large crystals. Fast cooling forms tiny crystals or no crystals.

Look for
Interlocking grains with <u>angular</u> shapes.

Sedimentary Rock

What It Is
Sediments (fragments of rock— silt, sand, and clay) accumulate, get compacted and cemented into rock.

Look for
<u>Rounded</u> grains, often layered. Chips off easily.

Metamorphic Rock

What It Is
Previously formed rocks undergo intense heat and pressure making them more compact and banded.

Look for
Sheet-like texture, flattened minerals in bands. Rock is very hard.

LHS GEMS: *Stories in Stone*
© 1995 by the Regents of the University of California

Observation and Display of "Mystery Rock"

Name of Our Group

Observations

[PLACE SAMPLE HERE]

Other Geologists' Opinions (√):

- Mineral
- Igneous
- Sedimentary
- Metamorphic

What type of sample is this?
Here is our group's opinion:

(circle one)

Igneous

Sedimentary

Metamorphic

Mineral

ROCK & ROLL RIDDLES

Write the name and number of the sample that goes with each "autobiography."

Rocks can't talk, but if they could, maybe they'd rhyme, perhaps they would
Talk in riddles to help us see, how they formed and came to be,
Then we could play a matching game to match their properties to their names.

A.
I was hot mushy magma movin' up on my own
Till I came to rest beneath some solid stone
Then I cooled for numerous centuries
Black and white crystals grew inside of me
The rock above me eroded in time
Till at last to the top I managed to climb.

I am _____   Number _____

B.
The elements that now are part of me
Once were in other rocks, but geologically,
They were swimming in water and came my way
Down deep in the Earth one molten day
It got so hot it propelled me higher
Pushed up into crevices—that cooled my fire
My elements crystallized in silvery hue
I am grayish and dense and shiny too.

I am _____   Number _____

C.
I used to have layers, dark and thin
For thousands of years, I lay within
Beneath a large pile of heavy clay
Soil and water that pressed away
I was squeezed and squeezed, and heated too,
My layers were baked through and through,
I got pushed straight up, tilted all around,
And somebody found me on mountainous ground.

I am _____   Number _____

D.
I used to a be hot, liquid-solid mix
Deep inside the Earth I did my tricks
Until I was erupted in volcanic fireball
Cooled very quickly, so my crystals are small,
I am dark, with tiny grains, to see them is tough,
I've got mineral pockets, and I feel sorta rough.

I am _____   Number _____

E.
I may look like I've been melted but I never did that
River leading to lake is where I first was at
Tumbled 'round, broken down, sand and pebbles I became
Then more rocks and pebbles over me did the same
They squeezed down upon all the pieces of me
Cementing them together quite solidly
Long ago the river dried, Earth's crust pushed me up
To this hill where I landed in a rock collector's cup.

I am _____   Number _____

F.

I used to be simply sedimentary
But then all kinds of stuff piled on top of me
I got pulled, I got squeezed, and heated besides
Till in layers my flattened minerals settled and crystallized,
I have shiny crystals, and if truth be told,
Of many kinds of rocks, I'm among the most old,
Erosion from above and pushing from below
Brought me up to the top for you to know.

I am _____   Number _____

• •

G.

When the seawater dried up,
It left me high and dry
My crystals are in cubes,
I'm quite a tasty guy
I once asked a chemist
What I'm made of, she replied:
"Mostly, I would say, you're sodium chloride."
If you cannot guess me, it's really not my fault
And by the way I have another name, which simply is ... _____ .

I am _____   Number _____

• •

H.

When water with lots of silica
Got trapped in a hole in the Earth
It cooled very very slowly
And that led to my birth.
So my crystals are very large, and besides,
They're shaped like prisms, with six shiny sides.

I am _____   Number _____

• •

I.

I was sculpted by volcanic energy
Lava pushed up fast and cooled so quickly
That no crystals were formed on front or back
A glass-like substance, shiny and black,
Treasured by many, truly unique
A special sheen that many seek
The shine I possess seems to come from within
By now I'm sure you've guessed it,
I'm _____ .   Number _____

• •

J.

There I was at the bottom of a lake
Water and sediment pressed till I ached
Pressure cemented my layers together
Pushed up to the surface I faced the weather
I'm grayish, brittle layers, with particles fine
Which name do you think just might be mine?

I am _____   Number _____

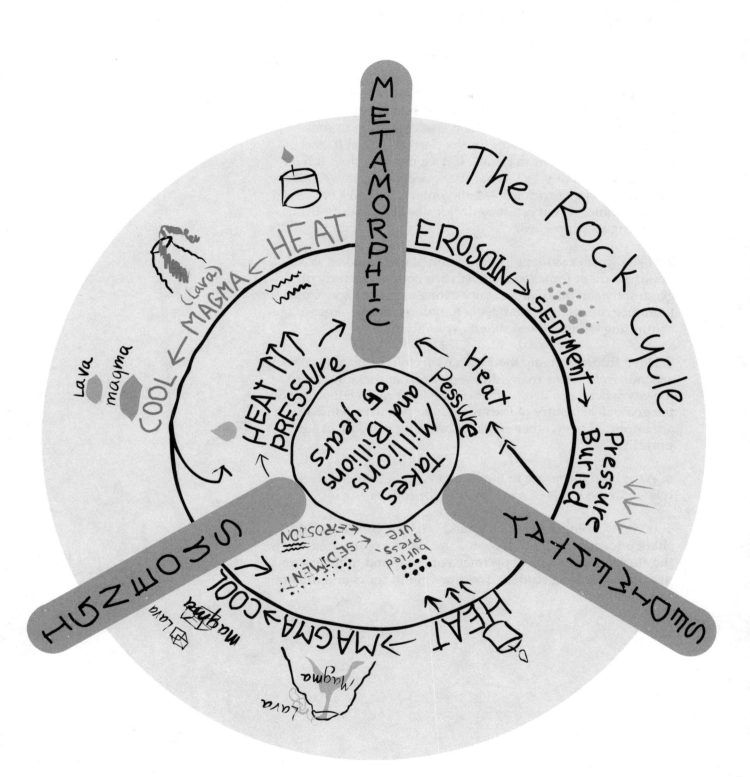

The Rock Cycle

Going Further (for the entire unit)

1. Invite your students to stay on the lookout for interesting rocks around their homes and neighborhoods, and those they spot when they go on trips. When students bring in rock samples, take a few minutes for the class to look at it, and decide—Is it a rock or mineral? If it's a mineral, does it have crystals that we can see? What shape are the crystals? If it's a rock, what type of rock is it—sedimentary, igneous, or metamorphic? What can they decide about its history from observing its properties?

2. Foster further student investigation of earthquakes, volcanoes, and other topics in the Earth Sciences. The study of geography can make great connections with geology! Consider literature connections that deepen understanding. Encourage continuing student interest and exploration!

3. Have students create their own rock collections of local rocks and minerals. How many different kinds of rocks are there in the city where you live? What do the local rocks tell you about the geological history of the area where you live? How many are imported from other areas for building projects, or for gravel?

4. What land forms are in the local area, and how are these related to the kinds of rocks that can be found?

5. Invite a geologist or "rock hound" to visit your class and share information about the geologic history of the area, show the students interesting specimens of rocks and minerals, and discuss continuing student questions about rocks and minerals.

Resources

Books

Crystal & Gem, Eyewitness Books, by Dr. R.F. Symes and Dr. R.R. Harding, Alfred A. Knopf, New York, 1991. This companion volume to *Rocks & Minerals* (see below) is also beautifully illustrated, with numerous large color photographs.

Earth Facts by Cally Hall and Scarlett O'hara, Dorling Kindersley, London, 1995. A convenient and nicely illustrated "pockets full of knowledge" book.

The Earth, Life Nature Library by Arthur Beiser, Time, Inc., New York, 1963.

A First Look at Rocks by Millicent E. Selsam and Joyce Hunt, illustrated by Harriett Springer, Walker Publishing Company, New York, 1984.

A Golden Guide Rocks and Minerals: A Guide to Familiar Minerals, Gems, Ores, and Rocks by Herbert S. Zim and Paul R. Shaffer, illustrated by Raymond Perlman, Golden Press, New York, 1957.

Matter, Life Science Library by Ralph E. Lapp, Time, Inc., New York, 1965.

Rock All Around by Margaret Farrington Bartlett, illustrated by John Kaufmann, Coward-McCann, New York, 1970.

Rocks & Minerals, Eyewitness Books, by Dr. R.F. Symes and the staff of the Natural History Museum, London, Alfred A. Knopf, New York, 1988. Lavishly illustrated, with numerous color photographs of rocks and minerals.

Rocks, Rivers, and the Changing Earth: A First Book About Geology by Herman Schneider and Nina Schneider, with illustrations by *Edwin* Herron, William R. Scott, Inc., Publishers, New York, 1952. See page 117 for a review of this excellent book.

Rocks Digest magazine describes itself as a publication that "informs and entertains students about the physical world of rocks." It is published six times a year and carries current geological, environmental, and earth science events and news. For subscription information contact *Rocks Digest* at P.O. Box 8186, Missoula, Montana 59807-8186.

Simon and Schuster's Guide to Rocks and Minerals, Edited by Martin Prinz, George Harlow and Joseph Peters, a Fireside Book published by Simon and Schuster, Inc., New York, 1978. A good introduction to rocks and minerals, with over 100 color photos.

Simon and Schuster's Guide to Gems and Precious Stones, by Curzio Cipriani and Alessandro Borelli, U.S. Editor Kennie Lyman, Translated by Valerie Palmer, a Fireside Book published by Simon and Schuster, Inc., New York, 1986. Nicely illustrated, this book includes over 100 gems, in natural and cut form.

Other Educational Resources

As noted on page 7, it is very useful to consult local rock shops and other geological resources and individuals with expertise in your locality.

Frey Scientific, 905 Hickory Lane, P.O. Box 8101, Mansfield, Ohio 44901-8101,(800) 225-FREY. This scientific supply company is a source for rock samples, salol, and other resource and educational materials.

Nystrom, a Division of Herff Jones, Inc., An Employee-Owned Company, 3333 Elston Avenue, Chicago, Illinois 60618-5898, (800) 621-8086. Fax: (312) 463-0515. This company offers many excellent educational resources, including a set of eight Earth Science charts with a 96-page teacher's guide, and a videodisc entitled "Understanding Earth." Their videodisc on "Evolution" also contains a number of earth science topics. Their catalog explains their offerings in colorful detail.

Scott Resources, Earth Science, P.O. Box 2121Q, Fort Collins, Colorado, 80522, (303)484-7445; 1 (800) 289-9299. This company offers a number of rock and mineral sample kits, equipment kits for testing various properties, earth science videolabs on a number of key topics (including the rock cycle), charts, and other educational resources. The "Physical Geography" section of their Earth Science catalog includes hands-on landform models and demonstration kits, and an 18-minute video program with teacher's guide on plate tectonics for Grades 7–12.

Ward's Natural Science Establishment, Incorporated, Post Office Box 92912, Rochester, New York 14692-9012, 1 (800) 962-2660. Toll-free fax: 1 (800) 635-8439. See especially *Ward's Earth Science Catalog*. When this GEMS guide was tested in classrooms nationwide, the class sets of rock and mineral samples provided were obtained from Ward's and the photographs on the inside back cover of this guide are reproduced with permission from the Ward's catalog. In addition to Ward's as an excellent source for the class samples, salol, and other related materials, the catalog itself has numerous color photographs of many rocks, minerals, and geological formations, and features audio visual materials on related topics in the earth sciences.

> *The Full Option Science System (FOSS), developed at the Lawrence Hall of Science and distributed by Encyclopaedia Brittanica Educational Corporation, 310 South Michigan Avenue, Chicago, Illinois 60604, (800) 554-9862, includes a number of activities in the earth sciences. For Grades 3 and 4, the "Earth Materials" module, with four main activities, brings students in touch with the basic building materials from which the Earth is made, and provides experience with simulated and real rocks, minerals, and techniques used by geologists to take apart and identify several important rocks and minerals. For Grades 5 and 6, the FOSS "Landforms Module." with five activities, develops concepts of physical geography and graphing. Students use stream tables and make topographic maps. In Activity 4, "Build A Mountain," students assemble a foam model of Mount Shasta, and create a topographic map and profile of it.*

You Are Part of the Earth's Story

There are many fine books on geology and the earth sciences. Several modern books that provide up-to-date information about tectonic plates are listed in the "Resources" section. One book for young people, although published more than 40 years ago, emphasizes essential lessons in memorable ways. If you should happen to be able to obtain it, we recommend it. The book is *Rocks, Rivers, and the Changing Earth: A First Book About Geology* by Herman Schneider and Nina Schneider, with illustrations by Edwin Herron, William R. Scott, Inc., Publishers, New York, 1952. In one passage, the reader is asked to imagine taking a walk in the countryside and picking up a pebble:

The pebble…has traveled far and wide. Long ago it was a drop of magma, molten rock that poured out deep from inside the earth. Perhaps when the magma cooled it formed part of a mountain that was later worn down and carried away by a rushing stream. Or the pebble may have been carried thousand of miles by a slowly moving glacier that finally melted and left it there for you to pick up. It has traveled to many places and has been part of many things…Even in a handful of soil there are bits of mountains, traces of sea animals, crumbled remains of last year's leaves, and the minerals of plants that lived millions of years ago…You are part of the earth's story. In your blood is iron from plants that drew it out of the soil. Your teeth and bones were once coral of the sea and tiny, beautiful sea animals. The water you drink has been in clouds high over the highest mountain in Asia and in lovely, misty waterfalls in Africa. The air you breathe has blown and swirled through places of the earth that no one has ever seen. Every bit of you is a bit of the earth, and has been on many strange and wonderful journeys over countless millions of years…

Rocks, Rivers, and the Changing Earth provides a very direct sense of the rock cycle, constant geological change, and the conservation of matter.

Rock is everywhere in the world, and everywhere it is being changed. High up on the mountainsides, rock is split by the heat of the sun and cracked apart by freezing water. It is rolled and crushed by rivers of water and rivers of ice. It is crumbled by lichens and dissolved by flowing water. Over countless centuries, rock is turned into soil everywhere in the world…But the journey of the rock is not ended. It is never ended. In every tiny part of every living thing are minerals that once were rock that turned to soil. These minerals were drawn out of the soil by plant roots, and the plant used them to form leaves and stems, flowers and fruits. When the plant was eaten by an animal, these same minerals became part of the animal. And still the journey of the rock is not ended, for nothing in the world remains unchanged forever. A rock is not always a rock; a rose is not a rose forever…The things of this world are formed again and again, out of the same materials of the earth. Nothing is lost. Over and over, these earth materials have been part of many things in many places.

Perhaps the most compelling strength of this book is the way it succeeds in connecting the study of rocks and minerals to you, the reader:

…The fresh crisp apple that you may eat today is as old as the hills. And when you eat it, a tiny bit of those hills becomes part of you… Just think of what that apple may have been before it became part of you! Once it may have been in the autumn leaves that fell and crumbled into the soil near the trunk of the apple tree. Years before it may have been in the shell of a robin's egg. And once it may have been part of a stalactite in some dark underground cavern. Perhaps for a short while it sailed high over the earth in a butterfly's wing. Long ago, it may have been in a kernel of corn planted by an Indian…Today, when you eat the apple, these parts of the earth become part of you who are part of the world.

Behind the Scenes

The following information may be helpful to you as you present *Stories in Stone*. It is **not** meant to be read out loud to students or copied for them, but to provide concise (and obviously partial) background on some of the main questions that you and your students may have. As we developed and tested this guide, these are some of the questions that arose. Many teachers and schools have excellent resource books on geology and related subjects, and we encourage you and your students to consult these as you pursue your curiosities about the Earth's crust. Several excellent books are listed in the "Resources" section of this guide.

What is Geology?

Geology is the branch of science devoted to investigation of the materials that compose the Earth, and to understanding the processes responsible for the formation of these materials. Within the general field of geology, there are various subfields, including mineralogy (which deals with all aspect of minerals) and petrology (which deals with all aspects of rocks).

Before the emergence of geology as a separate, integrated field during the mid- to late eighteenth century, miners and others involved in the extraction and processing of metals, clays, coal, and salts from the Earth had, in the course of their work, gained much practical geological information. Some philosophers had also formulated various geological theories, but many of these were unfortunately quite unrelated to this practical knowledge!

With the refinement of scientific methodology, which also occurred around this time, came more accurate study of the materials of the Earth and their processes of formation and change. During the last century, geology has truly come into its own. On the next several pages, we've attempted to summarize several major ideas in modern geology.

A more recent term that includes, but is not limited to, geology, is earth science, or the earth sciences. While this certainly includes the study of rocks, minerals, and the processes which form them, it can be extended to numerous related fields, such as environmental and atmospheric aspects of the Earth, oceanography, and even astronomy. The terms *geology* and *earth science* are often used interchangeably, however, and in this guide we may sometimes refer to geologists, sometimes to earth scientists.

How has modern geological thought evolved?

Four important concepts and/or scientific understandings evolved during the late eighteenth century and early nineteenth century to provide the foundation for modern geological thought. These include:

1. Determination that the Earth was of a much greater age than traditional estimates, which were often based upon interpretations of religious or folkloric texts, and which also assumed that the Earth was created just as we see it today. Until the early nineteenth century, most people believed that the Earth was only some thousands of years old! With the publication of *Principles of Geology* by Charles Lyell, the idea that the Earth was in fact many millions of years old began to gain wider acceptance. Lyell's work was distinguished by its descriptions of rock layer patterns, including locations of fossils, and by the logic used in drawing inferences from his observations. His work in turn was used by Darwin to strengthen his theory of evolution (see #4 below).

2. Classification of rocks according to the minerals they contained. As you and your students learn during this unit, classification is one of the central features of geology. The distinction between rocks and minerals that students learn, as well as the further classification of rocks according to the processes that create them, are important components of modern geological classification. In addition, analysis of crystalline structure, as well as chemical composition, can also be key factors. Many other properties and attributes are also taken into account. As the guide notes, when confronted with any one sample, classification can sometimes be quite a complex undertaking!

3. Recognition of the roles played by both the action of water and volcanic activity in the formation of rock layers observed on the Earth's surface. As the environmental "stories in stone" in this unit indicate, the heat associated with volcanic action gives rise to igneous rocks, and may also play a part in the formation of metamorphic rocks. Meanwhile, erosion, sedimentation, and other interactions with water also contribute to many geological processes. In a larger sense, both volcanic activity and the water cycle in all its forms play a major role in creating, shaping, and constantly transforming the crust of the Earth, from eruptions that build volcanic ranges, to floods that make oceans where there once was land.

4. The assumption that geological forces seen in operation at present should be used to explain the past history of the Earth. This principle is known as "uniformitarianism." By the mid-nineteenth century, the science of geology had greatly matured, due to the bringing together of scattered geological facts, and the refinement of this principle, which held that *all forces of nature have always been the same as they are now.* This concept, considered one of the cornerstones of all modern scientific thought, was further developed during the late nineteenth century by its most notable representative, Charles Darwin. It is interesting to note that Darwin's early work was in the field of geology, and he stated that the study of geology helped lead him toward his famous Theory of Evolution.

What is Plate Tectonics?

The development of these four main ideas, along with increasingly advanced and accurate geological field studies and rapid leaps in technology and science in general, set the stage for a new, over-arching approach. In the early part of the twentieth century the kernel of the Theory of Plate Tectonics was conceived, but it was not until quite recently that it has become widely accepted and more fully and accurately elaborated.

Note: The word tectonics *derives from the Greek, and pertains to construction, structure, and building. In geology, tectonics refers to the structures, forces, and conditions within the Earth that cause movements of the crust. Plate tectonics posits that large plates and their interaction provide the structural basis for movements of the crust.*

This important theory provides a single, all-encompassing concept to interpret many of the major geological features of the Earth. The theory holds that the entire Earth's crust is composed of about a dozen large, distinct, rigid pieces, or plates, which move over interior layers of the Earth. These plates are thought to move a few centimeters (about an inch) a year. (See illustration, page 132.) The dynamic interaction of these plates is seen as responsible for the formation of many large-scale features and zones of activity, such as mountain belts and ranges, volcanic and seismic (earthquake) activity, and other geological phenomena and events. Your students are introduced to this basic idea in Session 7 of *Stories in Stone*.

Although many people in the past may have noticed that the coastlines of the continents resemble pieces of a giant jigsaw puzzle, everyone supposed this was coincidence, since continents were seen as immovable. In 1915 Alfred Wegener audaciously hypothesized that the continents had in fact once been merged and had since moved. He also gathered evidence to show that if continents were fitted back together, the mountain ranges and rock types also often matched. His ideas, sometimes called "continental drift" were ignored, dismissed, or ridiculed. Opponents argued that there was no known mechanism for this to happen, no forces known that could move continents. It was not until the 1960s that enough evidence from a number of fields accumulated for most earth scientists to conclude that Wegener's hypothesis of continental movement was essentially correct, even though much of his reasoning and proposed evidence were flawed or inaccurate. From their experience in this unit, your students may be interested in finding out more about the Theory of Plate Tectonics.

What's really inside the Earth?

Humans have always wondered about and sought ways to find out more about the internal structure of the Earth. The records of a large number of ancient civilizations contain geological observations, including speculation about the inside of the Earth. From ancient myths to Jules Verne's science fiction journey to the center of the Earth, this has been a fascinating subject.

Over the past sixty years or so, a more well-rounded and widely accepted understanding of the Earth's internal structure has emerged. This is based on the combined evidence from many studies of Earth's physical characteristics. These studies have involved, for example, the properties of rocks and minerals, the densities of various materials, the Earth's magnetic field, and the transmission of seismic (earthquake) waves.

The modern view holds that the materials of which the Earth is made are separated according to their density into different zones. Heavy material is concentrated nearer the center, and lighter material nearer the surface. The three main Earth layers (sometimes also called units) are:

(1) the core, composed predominantly of iron and nickel;

(2) the mantle, composed of silicate minerals rich in iron and magnesium; and

(3) the crust, composed of rocks and minerals exposed near the surface.

The core of the Earth is thought to be a central mass about 7,000 kilometers in diameter. Though it makes up only about 16% of the Earth's volume, it is believed to account for 31% of the Earth's mass. It is thought that the core consists of two parts—a solid inner core and a liquid outer core. Rotation of the Earth is thought to cause the liquid core to circulate, and its circulation in turn generates the Earth's magnetic field.

The mantle surrounds or covers the core. It is estimated to be over 3,000 kilometers thick, and to account for more than 80% of the Earth's volume and close to 70% of its mass. The upper portion of the mantle, known as the asthenosphere, is believed to be a structurally weak, partially molten zone that is capable of flow. Movement within this zone is thought to be responsible for the volcanic and seismic activity at the Earth's surface.

The crust (or lithosphere) is defined as the outermost shell of the Earth. Embedded within it are oceanic and continental plates. A rigid, solid, strong layer about 100 kilometers thick, the lithosphere rests atop the more structurally weak, partially molten material of the asthenosphere. Rocks and minerals are the main materials in the Earth's crust.

Several of the books listed in the "Literature Connections" contain cleverly presented information for your students on what is thought to be inside the Earth.

What is a mineral? What is a rock?

In this unit, students learn a basic distinction between rocks and minerals— minerals are made of one substance, while rocks are most often assemblages of more than one mineral. There are some rocks, such as pure quartzite, that are made up of only one mineral, but most rocks are composed of two or more minerals. Minerals are the ingredients of which rocks are made. They are the building blocks of rocks.

Minerals and rocks are the most abundant inorganic (non-living) solid materials found on or near the Earth's surface. Everywhere we look in daily life and the natural world we find minerals and rocks, as well as products and substances derived from them. Nonetheless, coming up with succinct yet complete definitions of minerals and rocks can be challenging. It is difficult, for example, to make sweeping generalizations about the physical appearances of all rocks and minerals, due to the multiplicity of their forms and structures, as well as the highly individualized ways they are created, molded, and transformed. For this reason, geologists seeking to carefully define minerals and rocks look beneath the surface to underlying chemical and atomic structure.

The definitions below reflect generally accepted, basic notions regarding minerals and rocks: To find more detailed information or pursue further questions, consult well-illustrated resources, invite a geologist to talk to the class, or visit a well-stocked rock shop.

Thousands of different kinds of materials in the Earth's crust are termed minerals because they share the following characteristics: A mineral is a naturally-occurring solid with a definite chemical composition and an ordered arrangement of atoms. Minerals are homogenous, that is, composed of one substance throughout. The processes that form minerals are inorganic—not directly related to plant or animal life.

One of the most important external features of minerals is the regularity of their crystal faces. Some rocks are classified as to whether or not they contain a specific mineral—such a mineral is termed a "rock-forming mineral."

Rocks are assemblages of minerals. Usually there are numerous minerals in a given rock. Because of the immense number of ways in which the thousands of identified Earth minerals may be combined, rocks are much more difficult to define by their properties than minerals. Instead, as students learn in this unit, rocks are classified into three main categories (igneous, sedimentary, and metamorphic) based on their processes of formation.

What is a crystal?

Crystals are familiar objects, prized for their shapes, for their beautiful, multi-faceted reflections of light, and thought by some to possess spiritual and healing qualities. Your students gain direct experience with crystals in this unit, forming them through evaporation and through the cooling of a substance (salol) that represents magma, simulating the way crystals are formed from volcanic activity and cooling in the Earth. Distinctive crystal shapes are a defining feature of many minerals. Naturally, rocks exhibit crystals as well. Of course, as rocks are subjected to heat, pressure, or other forces over time, their crystalline features are also modified.

Scientifically speaking, crystals are defined by their structure. A crystal is an homogeneous solid with a regular geometric form, whose boundaries are naturally-formed smooth, planar surfaces (also called faces). This regular geometric form reflects an orderly three-dimensional internal structure. While minerals are *only* formed from non-organic processes (as defined above), crystals can be formed from **either** inorganic or organic compounds. Therefore, while many minerals are found in crystal form, not all crystals are minerals.

Study of the nature of matter and mathematical analysis have revealed an interesting "facet" of crystal formation—there are only a limited number of crystal shapes possible. All crystals, whether composed of organic or inorganic compounds, can be placed into seven general classes of symmetry, known as "crystal systems." Within these seven main crystal systems is a further sub-grouping of 32 "crystal classes." Mathematical analysis has shown that there are only 32 different ways of arranging atoms around a point—only 32 ways a crystal can be constructed so that it both obeys the laws of symmetry and is a three-dimensional solid (fills space in three dimensions). Consult geology resource books for additional descriptions and information on these seven systems and 32 classes of crystals.

How do minerals form?

Minerals are formed from solutions, melts, and sometimes vapors. In these generally disordered states, the atoms have a random distribution and orientation. But with changing temperature, pressure, or concentration, the atoms may combine to form an ordered arrangement.

Minerals form when a solution becomes supersaturated (that is, too enriched in dissolved compounds to hold them any longer), when a liquid solidifies, or when a vapor condenses. Therefore, temperature, pressure, and availability of chemical elements are three of the most important environmental factors affecting mineral growth.

Sodium chloride (halite) dissolved in an aqueous (water-based) fluid is an example of mineral formation from a solution. As the solution evaporates, the concentration of sodium and chlorine atoms within it increases to such an extent that these elements begin to group themselves together. Soon a point is reached where the remaining solution cannot retain in solution all of the resulting sodium chloride molecules and halite crystals begin to precipitate, or "drop out." When your students make salt crystals through evaporation, they are gaining experience with mineral formation from a solution.

Minerals are formed from melts in a similar way. The best example of this process is the formation of igneous rocks from molten magmas (batches of hot, liquid rock). When a given magma cools sufficiently, the atoms contained within it (which were free to move in any direction in a liquid state) are first attracted to each other and then arrange themselves in a definite order—forming nuclei of different minerals. With further cooling, these minerals grow larger and eventually form one solid mass. The salol activity in Session 4 *represents* this crystallization process, but, because salol is not a mineral (see below) the activity is a good representation of mineral formation, but not an actual case.

Although mineral crystallization from a vapor is less common than from a solution or melt, the underlying principles of the process are much the same. As the vapor cools, dissociated atoms, or molecules, are brought closer together, eventually locking themselves into a crystalline solid.

Is salol a mineral? What is it used for?

Salol, more technically known as phenyl salicylate, is an organic compound ($C_6H_4OHCOOC_6H_5$) which is produced **synthetically**, that is, through processes that do not occur in nature. Since minerals are defined as inorganic solids that occur naturally, salol is not a mineral.

However—just like natural mineral crystals—salol crystals also possess regular, geometric form and structure, resulting from three-dimensional internal order.

Salol also has other properties that make it an excellent choice for experiments like those students do in Session 4. Due to its relatively low melting point (108° F) and generally safe nature, salol is a substance that is often used to illustrate fairly rapid crystal formation. In addition, salol is used in the manufacture of various plastics, lacquers, waxes, polishes, sun tan oils, and creams.

What are gems?

Gems are very clear, pure, crystalline forms of a number of minerals that occur in nature. The main characteristics of gems that contribute to their value and appeal are: hardness, durability, rarity, color, transparency, and beauty. The few minerals which rank high in all of these gem characteristics tend to be the most highly prized gems. These, generally classified as "precious," include diamonds, rubies, emeralds, and sapphires. All other lesser-prized gems are classified as "semi-precious." Corundum, garnet, opal, topaz, tourmaline, zircon, and various varieties of quartz are among the more popular semi-precious gems.

Interestingly, many of the popular names of gems go back many centuries and were in use long before the science of geology developed. Modern names given to minerals usually end in "ite," such as flourite or malachite, but the majority of names for precious and semi-precious gems predate modern classification schemes, so we know and still call them by their historical names. Many names, such as agate, amethyst, diamond, and emerald are derived from the Greek language. Beryl is also derived from Greek but probably originally came from Sanskrit. In Greek, amethyst means "without drunkenness" and the stone was regarded as a remedy for intoxication. The names opal and sapphire are derived from Sanskrit, garnet and ruby from Latin, jade from Spanish, turquoise from French, and quartz from German.

The use of these historical names often means that their classification lacks the precision and accuracy of more scientific terminology. As a further complication, gem minerals possess many different varieties to which distinctive names have also been given. For example, over 20 different varieties of the common mineral, quartz, have been used for gem-related purposes. On the other hand, the semi-precious stone termed garnet by jewelers actually occurs in six distinct varieties—according to standard geological classification. Ruby and sapphire are differently-colored varieties of the same mineral—corundum. Emerald and aquamarine are both varieties of beryl. The history, naming of, uses of, and changing economic status of gems would make a very interesting special research project. Several books that might be helpful are listed in the "Resources" section.

How are rocks classified?

Many hundreds of different varieties of rocks are known to exist within the Earth's crust. However, all of these rocks can be generally grouped according to their origin. Three basic groups (or classes) are commonly described: igneous rocks, sedimentary rocks, and metamorphic rocks.

Igneous rocks, believed to comprise close to 95% of the upper 10 miles of the Earth's crust, are those that form when molten rock, or magma, cools and solidifies. If the cooling magma solidifies *before* reaching the Earth's surface, it forms **intrusive** igneous rocks. If the magma travels all the way to the surface before cooling it is called *lava*; when the lava solidifies it forms **extrusive** igneous rocks.

Notable characteristic features of igneous rocks include considerable hardness, a highly crystalline texture with a fabric of interlocking, angular mineral grains, a glassy or sponge-like appearance, and a general lack of layering or banding.

Sedimentary rocks are composed of materials (sediments) derived from weathering of previously-existing rock masses. These sediments usually accumulate as a result of the action of water, and later become compacted and/or cemented. Sediments fall into two basic categories: clastic (or detrital) sediments, from mechanical accumulation of individual grains; and chemical (or biochemical) sediments, produced by chemical precipitation from solutions. It's estimated that sedimentary rocks comprise about 5% of the upper 10 miles of the Earth's crust.

Sedimentary rocks exhibit a parallel arrangement of minerals which form layers (sometimes called beds). These layers are distinguished from each other by various differences, such as thickness, size of grains, or color. Students gain insight into this layering in Session 5.

Metamorphic rocks comprise less than one percent of all rocks found within the upper 10 miles of the Earth's crust. They are the most complex group. Derived from pre-existing sedimentary, igneous, or already metamorphic rock masses, metamorphic rocks have undergone mineral, structural, or textural changes. These changes usually result from exposure to large increases in pressure and temperature. Generally, these changes take place in a solid state, although it can happen that solids exchange chemical elements with the involvement of relatively small amounts of a fluid (usually mostly water).

Three major groups of metamorphic rocks are commonly recognized: (1) regional (or dynamic) metamorphic rocks, formed as the result of exposure to significant increases in temperature or pressure—or both—on a regional scale (e.g., those associated with mountain building processes that affect areas a few hundred to thousands of miles in extent); (2) contact metamorphic rocks, formed through contact with relatively hot, intrusive magma bodies, causing effects that occur over a fairly limit*ed area*; and (3) replacement metamorphic rocks, formed when water, carrying dissolved minerals, exchanges one mineral in a rock for another it has in solution.

The most notable distinguishing feature of metamorphic rocks is their foliation, that is, parallel alignment of minerals along planar surfaces, or, in more descriptive terms, the splitting or arrangement of the rock into leaf- or sheet-like structures. Such an alignment of fine-grained minerals produces surfaces along which the rock has a tendency to split. This property is generally referred to as "slaty cleavage," which refers to the main characteristic of the metamorphic rock known as slate.

Parallel alignment of medium-grained, platy minerals, such as mica, result in some metamorphic rocks having a shiny, scaly appearance. Metamorphic rocks that exhibit this property are known as schists. More coarsely grained metamorphic rocks show less distinct foliation but a much broader banding of minerals. Such rocks are known as gneisses.

Where do the rock and mineral samples come from, how are they formed, what are they used for?

The following chart contains basic information about the rock and mineral samples used for the activities in this guide. The technical terms "felsic" and "mafic" are used—they are defined when first used (under Basalt, Composition). The chart is designed for quick reference, not as an authoritative source. For more information, consult standard geology reference works, such as those listed in the "Resources" section. If you select different class samples, you may want to gather some basic information about them.

Basalt

Composition: An igneous rock composed of approximately 50% felsic minerals (those which are light-colored and contain mainly aluminum, silica, calcium, and oxygen) and 50% mafic minerals (those which are dark-colored and contain mainly iron, magnesium, oxygen, and silica).

Origin: Cooling and solidification of mafic magmas on or near the Earth's surface.

Occurrence: The most abundant of all igneous rocks, found on the surface in volcanic regions throughout the world, particularly as massive lava flows. Noted deposits include those that form the Columbia Plateau of Oregon and Washington in the U.S., and the Deccan Plateau in western India.

Economic Use: Primarily used in road building. A vesicular form that develops near the top of lava flows, known as scoria, is also used in gardening and landscaping. (Vesicular means having small, spherical cavities, usually formed by gases or vapors.)

Conglomerate

Composition: A sedimentary rock composed of rounded pebbles and/or boulders cemented in a matrix of finer-grained material.

Origin: Consolidation of rounded extremely coarse-grained sediments larger than or equal in size to gravel (8 mm diameter).

Occurrence: Fairly common sedimentary rock found throughout the world in areas where ancient mouths of swift flowing rivers, alluvial fans, or deltas have become exposed. Characteristically thick-bedded to massive.

Economic Use: Of limited use as source material for concrete aggregates. Some tightly cemented varieties are capable of taking a good polish and have been used as ornamental stone.

Galena

Composition: A mineral composed of approximately 87% lead and 13% sulfur. Also may contain trace amounts of silver.

Origin: Precipitation resulting from cooling and evaporation of warm aqueous (mainly water) solutions containing numerous dissolved elements.

Occurrence: Most often found in veins long, tabular bodies that form when minerals fill pre-existing fissures or fractures alongside other sulfide minerals and usually quartz. Noted deposits are found in Freiberg, Saxony; Cornwall, England; and the Missouri, Kansas, and Oklahoma Tri-State district in the U.S.

Economic Use: Primary source of lead used in making pipes, shot, glass, solder and as shielding material against radioactive materials.

Granite

Composition: An igneous rock composed of approximately 60% potassium feldspar (potassium, aluminum, oxygen, and silica), 30% quartz (silica and oxygen), and 10% mafic minerals.

Origin: Slow cooling and solidification of felsic magmas at some depth within the Earth, usually in large bodies.

Occurrence: The best known of igneous rocks formed below the Earth's surface, granite is usually found in continental areas where the roots of ancient mountain ranges are now exposed. Prominent granite occurrences have been found in the Manitoba and Ontario regions of Canada, the central Alps of western Europe, and the Sierra Nevada range in the western United States.

Economic Use: Widely used for architectural construction, as ornamental stone (interior and exterior), and for monuments. Also, potassium-rich granite is crushed and used as fertilizer.

Halite

Composition: A mineral composed of approximately 60% chlorine and 40% sodium.

Origin: Precipitation resulting from evaporation of restricted or cut-off bodies of sea water.

Occurrence: A very common mineral, often occurring in extensive beds that formed by evaporation and drying-up of ancient, enclosed bodies of salt water. Extensive bedded deposits of halite or "rock salt" are widely distributed throughout the world. Important mining of salt occurs in China, Great Britain, Germany, Canada, and Italy. In the United States, the world's largest producer, salt is mined in fifteen states, including Ohio, Kansas, New Mexico, and New York.

Economic Use: Principal source of rock salt, used as a seasoning and for medicinal purposes.

Obsidian

Composition: An igneous rock that is actually an opaque, amorphous solid composed of tiny pockets of microscopic mafic minerals set in a glassy matrix. The chemical composition of obsidian, also known as volcanic glass, varies with location.

Origin: Extremely quick cooling and solidification of magmas, before many minerals can form.

Occurrence: Found in volcanic regions through out the world, primarily where felsic magmas similar in chemical composition to granite have fairly recently (within the pastone hundred thousand years or so) been erupted onto the surface. Noted deposits occur in California (Inyo, Imperial, and Modoc counties), Oregon (Crater Lake region), and Mexico (near Pachuca).

Economic Use: Ancient peoples valued obsidian highly, chipping and flaking it into knives, spearheads, and many other implements with razor sharp edges.

Quartz

Composition: A mineral composed of approximately 53% oxygen and 47% silica.

Origin: Slow cooling and solidification of felsic magmas, or precipitation from warm, aqueous solutions.

Occurrence: A common and abundant mineral occurring in a great variety of geological environments. As a rock-forming mineral, quartz is a major component of granite and other felsic igneous rocks, a very common constituent of sandstone (a sedimentary rock), and is usually present in a variety of metamorphic rocks, including schist, gneiss, and amphibolite. In addition, quartz is the most common mineral found in association with metal-bearing veins. Notable localities where large crystals are found include: the Alps; Minas Gerais, Brazil; the Malagasy Republic; and Japan. In the United States, the finest quartz crystals are found in Paterson, New Jersey; the Ouachita Mountains of Arkansas; the region surrounding Pike's Peak, Colorado; and Kern County, California.

Economic Use: Small crystals are used in the manufacture of electronic devices, while larger crystals are used as ornamental stones. Melted quartz is used in the making of lenses and prisms, as well as laboratory ware that is highly resistant to chemical action. Sand rich in tiny quartz crystals is used in making glass.

Schist

Composition: A medium- to coarse-grained metamorphic rock with a laminated appearance that exhibits well-developed foliation. (Foliation means the splitting up or arrangement of the rock in leaf-like structures.) This foliation results from parallel orientation of platy minerals such as mica and talc in thin, flat planes. Although the term schist is mainly a descriptive one, specific varieties are named after their most characteristic minerals (e.g., mica schist and chlorite schist).

Origin: Regional metamorphism of sedimentary, igneous, or other metamorphic rocks, under temperature and pressure conditions ranging from high to low.

Occurrence: A fairly common rock found in areas which have experienced low to high grade regional metamorphism (e.g., mountain belts). Notable locations where schists that contain sought-after minerals are found include: Alaska (Wrangall district), California (Marin, Mendocino, and Sonoma counties), Connecticut (Litchfield, Middlesex, and Norwich counties), Georgia (Fannin County), North Carolina (Cherokee County), and Ontario, Canada (Hastings County).

Economic Use: Some schists are minor source rocks for minerals of economic value, such as graphite schist for graphite. Some firm, erosion-resistant schists (e.g., quartz-mica schist) are used as building stones.

Shale

Composition: A fine-grained sedimentary rock composed of loosely consolidated clay minerals (those containing mainly aluminum, silica, hydrogen, and oxygen), small amounts of quartz, and some mica.

Origin: Compaction and consolidation of mud, silt, or clay deposited in the quiet environment of lake or ocean bottoms.

Occurrence: A very common sedimentary rock, normally found worldwide in massive, well-laminated layers that are associated with mesas, buttes, and similar cliff-like features. The environments within which most shales occur are seldom noted for the presence of valuable materials.

Economic Use: Used in the manufacture of bricks, cement, pottery, and other ceramic products. Also, oil-bearing shales represent a potential supply of fossil fuel.

Slate

Composition: A fine-grained metamorphic rock composed chiefly of well consolidated clay minerals, with some quartz and mica. Slates are distinguished by a property known as "slaty cleavage" which refers to their ability to be broken along a flat plane and split into thin, broad sheets.

Origin: Regional metamorphism of mudstones and shales under relatively low temperature and pressure conditions.

Occurrence: Slate is generally found in areas throughout the world which have experienced low grade metamorphism (under relatively low temperature and pressure conditions). This metamorphic rock usually occurs as steeply-tilted outcrops with jagged or irregular outlines, flaking, and separation along foliation surfaces caused by weathering. The environments within which slates occur are generally not noted for the presence of valuable materials (including minerals).

Economic Use: Primarily used in the manufacture of roofing and flooring materials, blackboards, and electrical switchboards.

Varved Shale

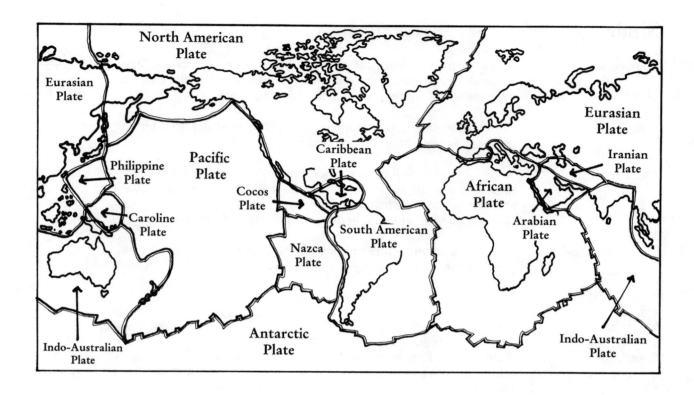

The surface of the Earth consists of about a dozen plates. Depending on how the exact boundaries are delineated, there are some differences of opinion concerning the exact number of plates. Some sources state that there are 15 plates (as shown above). Plates move toward, away from, or past one another at a rate of several centimeters a year. However, it's important to note that humans have only been closely observing and recording this movement for a limited time, so all generalizations must be qualified by recognizing our still very limited knowledge. **As shown in the drawing above, plates may include both continental lithosphere and oceanic lithosphere, so a single plate can include both land and sea.** Australia lies in the middle of a plate, while Iceland has a plate boundary passing right through it. Much research continues on tectonic plates and their rate of movement, relation to earthquakes and volcanic activity, and to many other phenomena.

Assessment Suggestions

Selected Student Outcomes

1. Students learn to distinguish between rocks and minerals, and to describe some of their properties.

2. Students improve their abilities to classify rocks as igneous, metamorphic, or sedimentary.

3. Students learn to recognize several crystal shapes, to identify them in rock and mineral samples, and to describe how some crystals form.

4. Students deepen their understanding of the processes by which rocks and minerals are formed, and how they can be transformed through the rock cycle.

5. Students gain insight into the processes that build up landforms (volcanism and mountain-building) and wear landforms down (weathering and erosion) and may be introduced to the theory of plate tectonics.

Built-In Assessment Activities

• **Classifying Rocks and Minerals.** In Session 1, students identify properties of and begin classifying ten rock and mineral samples. In Session 8, they make a final classification, based on information they've gained over the course of the unit about standard classification. By observing and taking notes during these two sessions, the teacher can gain insight into student progress in ability to identify important properties, distinguish between rocks and minerals, and classify rocks into the three major classes. (Outcomes 1, 2)

• **Observing Crystal Growth**. In Session 4, students conduct experiments comparing crystal growth at room temperature and at a colder temperature. They fill out data sheets that ask them to draw the crystals formed and describe in words how the shapes and sizes of the crystals differ. By circulating during this task and later reviewing the data sheets, teachers can assess student observation skills and find out if students have been able to infer the conditions under which large or small crystals tend to form. (Outcome 3)

• **Making Crystals**. At the end of Session 5, students are asked to compare the processes by which salt and salol crystals formed. Their responses can help teachers evaluate how well students understand some of the processes in the formation of minerals. Asking students to write their responses to this comparison provides a record of their understanding at this point. (Outcome 4)

• **Mystery Rocks**. In Session 8, students are asked to classify "mystery rocks" brought in during the unit. This activity is much more challenging than classifying the ten samples, as it asks students to apply their observational skills and general knowledge gained in the unit to an entirely new sample. This can help teachers assess students' abilities to distinguish rocks from minerals, characterize the three main types of rocks, and make inferences about how the mystery sample may have been formed. It can also provide an opportunity for students and teachers to become aware of the ambiguities in classification and the role of controversy and debate in science. (Outcomes 1, 2, 3, 4)

Additional Assessment Ideas

• Claymation Drama. Ask students to perform a "claymation drama" using colored clay to demonstrate: how landforms are made (mountains, cliffs, valleys); how they are worn down (eroded, weathered); and how larger geological events and processes, including plate tectonics, can change landforms and create different kinds of rocks and minerals. (Outcomes 4 and 5)

• Rock Cycle Diagrams. This "Going Further" at the end of Session 7 combines an overhead display with teacher-led discussion to provide both a review and an opportunity to question students about their understanding of and ability to explain processes of rock formation described during the clay modelling of Sessions 6 and 7. (Outcomes 4 and 5)

• The Story of Ms. Terry Rock. This open-ended activity, suggested in the "Going Further" for Session 7, encourages students to tell the "life story" of a rock. It provides an opportunity for students to demonstrate what they've learned about rock formation and geological change over time, in a creative context. (Outcomes 4, 5)

• Rock & Roll Riddles. This "Going Further" activity following Session 8, using activity sheets that follow, invites students to play a matching game—matching each poem with one of the ten samples. Asking students to write about or discuss why they matched the poems and samples as they did provides an opportunity to see how well they have grasped many basic concepts in the unit. (Outcomes 1–5)

• Our Town. As a class project, have students research local geological landforms and find out how they were created, sculpted, or in other ways affected by natural processes. This could include mountains, hills, valleys, rivers, lakes, plains, and so on. If possible, invite a local geologist or "rock hound" to speak with the class about the geological history of your area. Then, have each student devise a poster or diorama to artistically dramatize at least one aspect of this history and present it to the class. Outcomes 4, 5)

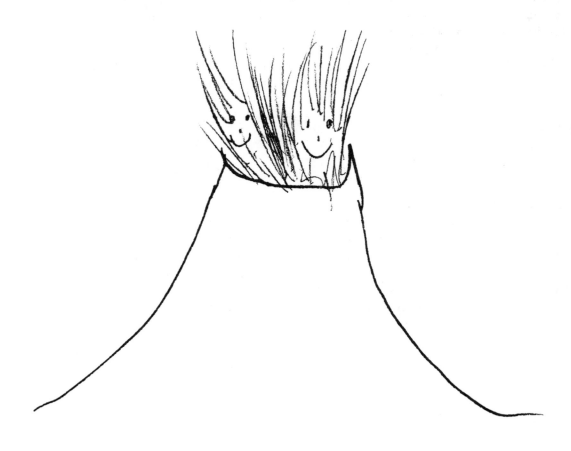

Literature Connections

Given that geology is a study of materials and processes that make up much of the world around us, it's natural that much literature contains brief references to rocks and minerals and/or background descriptions of geological settings. Books that might make a deeper and more meaningful connection include those whose plots hinge around a major geological event, or those which trace a rock or geological formation over time. A classic like Jules Verne's *Journey to the Center of the Earth*, although scientifically dated, is a great literature connection.

A number of the books listed here emphasize changing geological circumstances over time—this connects well to the underlying processes emphasized in the guide. Several books listed express the special cultural, even spiritual, role that rocks can play in our lives. We have not listed a large number of books that focus on volcanos, because that is not the emphasis of this particular GEMS guide. Many such books, however, could be good literature connections for the salol activity that models igneous rock formation, and to igneous rocks in general. Similarly, we have not listed books about earthquakes because that is not the guide's emphasis. Books about earthquakes could make good literature connections with the introduction of plate tectonics in Session 7.

Grade level estimates are purposely very flexible—this teacher's guide ranges from 4th through 9th grade, which is quite a spread! Also, sometimes a book apparently intended for a younger audience can be quite appropriate for older students when read in connection with a science or mathematics activity they've just explored. A book like *Everybody Needs A Rock*, for example, which at first glance seems for early grades, attains a universal and poetic appeal. We're sure you and your students have other favorites—send us your suggestions. We'd like to add them to the next edition of this guide and include them in the GEMS literature handbook, *Once Upon A GEMS Guide: Connecting Young People's Literature to Great Explorations in Math and Science*.

Bearstone
by Will Hobbs
Atheneum, New York. 1989
Avon Camelot, New York. 1991
Grades: 5–9

In this moving story, Cloyd, a Navajo boy searching for his path in life, finds an ancient turquoise stone carved in the shape of a bear. "He felt he was meant to cross the cliff and find this stone. He had earned this bear-stone; his grandmother would understand. She was the only person he knew who remembered the old ways and believed in their power." In addition to the stone having great significance for him, the book is also very descriptive of the geology of the Southwest. Students could watch for these descriptions and relate them to what they've learned about rock formation.

The Clay Marble
by Minfong Ho
Farrar Straus Giroux, New York. 1991
Grades: 7–adult

This novel of war-torn Cambodia in the early 1980s tells the story of a 12-year-old girl named Dara, her family, and her friend Jantu. Please be advised that the book is clearly intended for older students as it includes descriptions of killings and other grim realities of war and the dislocation of entire populations. It is primarily a powerful multicultural work and human statement, and its connection to this guide centers on the process of modeling things out of clay, including small clay spheres. Dara and Jantu also fashion small dolls and animals out of clay to create an entire miniature world—a world, unlike their own, filled with peace and happiness. The author was in Cambodia as a relief worker and received a clay marble as a gift from one of the children she met there.

Everybody Needs A Rock
by Byrd Baylor; illustrated by Peter Parnall
Aladdin Books/Macmillan Publishing Co., New York. 1985
Grades: all ages

What are the qualities to consider in selecting the perfect rock for play and pleasure? The properties of color, size, shape, texture, and smell are discussed in such an appealing way, you'll want to rush out and find a rock of your own. This poetic tribute is an inviting book for all ages.

Finding the Green Stone
by Alice Walker; illustrated by Catherine Deeter
Harcourt Brace Jovanovich, San Diego. 1991
Grades: 4–8

While this book, by a noted African-American author, does not connect directly to the scientific and geological information that students gain in thisGEMS guide, it tells an affirming story that connects the image of a shining green stone to that special something inside each one of us. There is strong emphasis on building self-esteem and living in harmony with our larger community. Beautifully illustrated, your students may especially enjoy realizing that the last picture in the book, portraying the Earth, is also a shining green stone.

How a Rock Came to Be In a Fence on a Road Near a Town
by Hy Ruchlis; illustrated by Mamoru Funai
Walker, New York. 1973
Grades: 2–5

This book traces the history of a gray rock from its origin as a pile of seashells to its present shape and location three million years later. The sense of geological change comes across clearly. Written for younger students, it could help strengthen an intuitive sense of the rock cycle concept that is explored in this GEMS guide.

How to Dig a Hole to the Other Side of the World
by Faith McNulty; illustrated by Marc Simont
HarperCollins, New York. 1990
Grades: K–8

A child takes an imaginary 8,000 mile journey through the Earth and is able to see what is thought to be inside. This book can further stimulate your students' natural curiosity about what's beneath the Earth's crust. The book begins with this advice for the digger: "Find a soft place. Take a shovel and start to dig a hole. The dirt you dig up is called loam. Loam, or topsoil, is made up of tiny bits of rock mixed with many other things, such as plants and worms that died and rotted long ago." The digging goes on, through the next layer of clay, gravel, or sand, through rockier soil, to bedrock. But it doesn't stop there! Diggers are instructed on how to get through bedrock, through the water table, through magma, the mantle, the outer core, the inner core, to the center of the Earth, using ingeniously imagined equipment that can withstand the temperature and pressure.

If You Are A Hunter of Fossils
by Byrd Baylor; illustrated by Peter Parnall
Aladdin Books/Macmillan Publishing Co., New York. 1984
Grades: 3–6

This book imagines a "hunter of fossils" finding signs of an ancient sea in the rocks of a west Texas mountain. It poetically describes how the area may have looked millions of years ago and what happened to cause it to change. Embedded in the text are many references to different kinds of rocks, geological events, and a strong sense of the interconnectedness of nature.

Iktomi and the Boulder
by Paul Goble
Orchard Books/Franklin Watts, New York. 1988
Grades: K–6

Iktomi (pronounced *eek-toe-me*) is the Sioux name for the clever and magical trickster who appears in a number of Native American cultures. Iktomi attempts to "defeat" a massive boulder, with the assistance of some bats. Your students make their own "boulders" as they work with clay in Sessions 6 and 7. Interestingly, this Iktomi story connects to the geological landscape—it "explains" why the Great Plains are covered with small stones.

Journey to the Center of the Earth
by Jules Verne
Available in many editions (see below)
Grades: 6 to adult

This science fiction classic sought to visualize an answer to a question people have wondered about for ages—what is inside the Earth? Equipped with what Verne imagined, based on the scientific information at that time, to be an adequate tunneling and exploratory craft, the expedition sets off. Although this book, like many Jules Verne works, has become somewhat dated as science learns more, it still makes an excellent connection. Students could investigate the ways in which Verne's descriptions differ from what scientists now believe to be the case. (Recent editions include Bantam Classics, 1991; Tor Books, 1992; Lake Publishing, 1994; Puffin Books,1994.)

The Lucky Stone
by Lucille Clifton; illustrated by Dale Payson
Delacorte Press, New York. 1979
Grades: 3–7

This book, by a leading African-American poet, traces a "lucky stone," shiny black as nighttime, that has been passed on, generation after generation, from slavery times to the present, and become a family talisman of freedom, protection, and love. Noteworthy for its sense of history, the geological connection in this book is to the special place that rocks can hold in our lives—some even become "lucky stones."

The Magic School Bus Inside the Earth
by Joanna Cole; illustrated by Bruce Degen
Scholastic, New York. 1987
Grades: 2–8

On a special field trip to the center of the earth, Ms. Frizzle's class learns firsthand about different kinds of rocks and the formation of the Earth and its structure. Presented in the informal and humorous style of this series, this book also succeeds in conveying plenty of solid scientific information. Although not primarily for older students, they may still be amused and educated! Reading it is a good way for interested students to begin learning more about geology and the forces at work inside the Earth.

Pablo Neruda Five Decades: A Selection (Poems 1925–1970)
by Pablo Neruda; English translations by Ben Belitt
Grove Press, New York. 1974

This bilingual edition contains selected poems of this world famous Chilean poet and Nobel prizewinner. The poems appear on facing pages in both Spanish and English, and include Las piedras del cielo (Skystones), a series of of more than 20 poems on rocks, minerals, and geologic processes. A number of selections from Skystones have been reprinted in this teacher's guide, with the kind permission of the publisher.

Paul's Volcano
by Beatrice Gormley; illustrated by Cat Bowman Smith
Houghton Mifflin, Boston. 1987
Grades: 5–8

When Adam and Robbie see Paul's science fair project, a model of a volcano (complete with smoke and eruption sound track), they make it the symbol of their new club. The "Vulcans" conduct rituals with the model volcano, chanting their password "Magma, Magma!" But what begins as a playful imitation of stereotypical legends about sacrificial volcano ceremonies turns into a series of unexplained and fearful events. Qualities of leadership and the meaning of accomplishment are explored as the strange events surge like lava down a mountainside. There is some scientific information throughout, including a description of the eruption of Mount St. Helens. In the end, the spirit of friendship triumphs over the evil genie of the volcano.

The Rock
by Peter Parnall
Macmillan, New York; Collier Macmillan Canada, Toronto; Maxwell Macmillan International Publishing Group, New York. 1991
Grades: 2–6

Over the years a rock provides shelter, a hiding place for animal and human hunters, and protects food and water sources—until it is struck by lightning. Then a white birch tree sprouts from the blackened rock, growing into a stand of trees. This book is an affirmative portrait of nature's multiplicity and endurance, emphasizing the solidity and permanence we associate with rock.

The Rock Hunters
by Lorraine Henriod; illustrated by Paul Frame
G. P. Putnam's Sons, New York. 1972
Grades: 2–6

Two boys help find an unusual petrified log while on a rock hunting expedition with the geology professor father of one of the boys. There is a discussion of plate tectonics: "These plates sometimes bump into each other. One plate may move underneath another one. Part of the earth is pushed up high, forming mountains. Other parts of the earth sink. Scientists are really detectives. We try to figure out from rocks how the world used to be." Those desperately seeking books with positive role models of scientists from diverse cultures will be pleased to see that the geology professor is African-American. While this book is geared for younger readers, it contains concise geological information on rock classification that aligns well with the activities in the GEMS guide. We did note one non-geological illustration error—an Eastern blue jay is depicted, although the text refers to a Canada jay.

I have already mentioned the fact that the Self is symbolized with special frequency in the form of stone, precious or otherwise...In many dreams, the nuclear center, the Self, also appears as a crystal. The mathematically precise arrangement of a crystal evokes in us the intuitive feeling that even in so-called "dead" matter, there is a spiritual ordering principle at work. Thus the crystal often symbolically stands for the union of extreme opposites—of matter and spirit... Many people cannot refrain from picking up stones of a slightly unusual color or shape and keeping them, without knowing why they do this. It is as if the stones held a living mystery that fascinates them...

from *Man and His Symbols*, by C.G. Jung, from the chapter
"The Process of Individuation" by M.-L. von Franz,
Dell Publishing Company, New York, 1964.

Summary Outlines

Session 1: Properties of Rocks and Minerals

Getting Ready
1. Before beginning the unit, obtain sets of rocks and minerals and prepare other materials.
2. Place samples and other materials on trays.
3. Gather other items listed in "What You Need."

Scratching the Earth's Surface
1. Discuss the Earth's crust briefly. If students wanted to find out more about it, what might they do?
2. Digging or drilling is difficult and expensive. Much can be learned by "scratching the surface" and examining samples.
3. Groups of four will examine samples.

Exploring Rocks and Minerals
1. Demonstrate use of magnifying lenses.
2. Distribute samples to each group.
3. First without, then with lens, have students examine, then discuss similarities and differences of samples.

Classifying Rocks and Minerals
1. Have students sort samples by a characteristic they choose.
2. Have other groups "guess the sort."
3. Discuss ways different groups chose to sort.

Introducing Properties
1. Define property as an observable characteristic that distinguishes one sample from another.
2. Discuss those (such as size and weight) that are less helpful in distinguishing one *kind* from another.
3. Explain that classifying is a good way to begin learning more about Earth's crust.
4. Ask if students know names of samples and point out key.
5. Collect samples and lenses.
6. As appropriate, discuss landforms of Earth's crust.
7. Ask students to bring in "mystery rocks."

Session 2: Distinguishing Rocks from Minerals

Getting Ready
1. Put about 2 teaspoons of kosher salt in each cup.
2. Have 3–5 cups of very hot water available.
3. Assemble materials for first part of session on trays; for second part on paper plates.

Distinguishing Rocks from Minerals

1. Ask students what they think is the difference between a rock and a mineral.
2. Distribute class sets of samples and lenses. Hold up samples of halite (# 2) and granite (# 3).
3. Have volunteers describe these samples. Take several responses. Emphasize that granite seems composed of many grains, halite of just one substance.
4. Have students arrange samples: on left, those composed of one substance; on right, of several substances; in middle, those difficult to classify this way.
5. Ask one group to report, listing results on chalkboard, and other groups to comment. Circle samples everyone agrees on.
6. Explain that samples made of one substance are minerals, those of more than one are called rocks.
7. Tell students it's sometimes hard to see individual mineral grains that are very small.
8. Explain that minerals are often in the form of regular geometric shapes called crystals. Have students pick out samples they think are crystals.
9. Collect samples and other materials.

Growing Salt Crystals

1. Tell students that crystals form inside the Earth and ask how might that happen.
2. Demonstrate procedure for dissolving salt in spoon/evaporation dish.
3. Have students, working in pairs, conduct the procedure and carefully place solutions in location to dry.
4. Ask for predictions of what might happen.
5. Remind students to bring in "mystery rocks."

Session 3: The Shapes of Mineral Crystals

Getting Ready

1. Copy crystal shape masters onto card stock.
2. Make the two crystal models students will do in class.
3. Assemble all needed materials.

Introducing Crystal Shapes

1. Encourage students to examine crystals grown in last session.
2. Review mineral, rock, and crystal definitions.
3. Shapes of crystals give mineral classification clues.

Constructing Crystal Models

1. Explain how to construct paper model of cube.
2. Have students construct both cube and hexagonal prism. Ask how many faces each shape has.

Comparing Crystal Models with Mineral Samples

1. Distribute class sets and magnifying lenses.
2. Ask if any of the samples have similar shapes to the paper models.
3. Help students identify halite and galena as having cubic shapes, and quartz as having an hexagonal prism shape.

Examining Previously Grown Mineral Crystals

1. Have a team member get previously grown crystals.
2. Ask them to observe closely, describe, and compare to samples.
3. Is there a mineral in the set that is similar? [yes, halite]
4. Explain that halite is a naturally-occurring salt crystal and ask where it might be formed and where table salt comes from.
5. Remind students to bring in "mystery rocks."

Session 4: Formation of Igneous Rocks

Getting Ready

1. Obtain salol.
2. Conduct the experimental procedure yourself.
3. Place materials for demonstration in appropriate place and assemble materials on trays.
4. Set aside class sets of samples and ice for later in session.

Introducing Igneous Rocks

1. Explain that one way to classify rocks is to study how they were formed.
2. Introduce igneous rocks (formed when magma cools and solidifies).
3. Some form when magma cools slowly inside the Earth; others when lava comes out of the Earth and cools more quickly.

Observing Crystal Formation at Room Temperature

1. Students first work in pairs to grow salol crystals at room temperature. Demonstrate the procedure, without lighting candle or melting salol.
2. Caution students to do experiments over the tray. Each pair should pour less than half of the salol crystals into their spoons (they will need left over for "seed crystals").
3. Introduce data sheet; each student is to complete one.
4. Distribute trays of materials.
5. Allow time for completion of experiment and drawing of crystals formed on the data sheet. Encourage discussion.
6. Ask students to describe how the crystals grew.
7. Explain that in the next part they will see what happens when the salol cools more quickly.

Comparing Crystal Formation at Different Temperatures

1. Give out ice cubes and paper towels. Students now work in teams of four. Students leave as is one spoon that formed crystals at room temperature. They remelt the crystals in the other spoon, then put bowl of that spoon on ice to cool more quickly, and add "seed crystals."
2. Students take turns comparing the crystals in the two spoons, with naked eye and magnifying lenses. Students draw crystals on data sheet and describe observed differences in writing.
3. If there is time, students could time how long it takes crystals to form at both room and colder temperatures by remelting in both spoons and simultaneously observing formation of one at room temperature and the other with ice.
4. Have groups blow out candles. Collect all materials.

Observing Igneous Rocks

1. Ask students to imagine the melted substance in spoons was volcanic magma. Help them articulate that larger crystals formed with slower cooling.
2. Ask how these findings might apply to igneous rocks in the Earth's crust—which ones would have the largest crystals? [The ones that cooled more slowly.]
3. Explain that magma cools more slowly inside the Earth, while lava that erupts at the surface of the Earth or under the ocean cools more quickly. Some lava cools so quickly that the igneous rocks formed have no crystals at all. Whether cooled slowly or quickly, the crystals in igneous rocks tend to have angular shapes.
4. Distribute the class set of samples. Ask them which rocks they think are igneous. [granite, basalt, obsidian]
5. Have students examine these three carefully and place them in order of how fast the magma cooled.
6. Discuss results and let students know that obsidian cools so fast there is no time for crystals to form.
7. Ask where on Earth igneous rocks would be formed.
8. Challenge older students to consider the difference between how the salt and salol crystals formed.

Session 5: Formation of Sedimentary Rocks

Getting Ready

1. Obtain sand, silt, and clay.
2. Use newspaper to cover student work areas.
3. Have a pitcher(s) with water available.
4. Choose outdoor location for students to collect soil samples.
5. Assemble materials on trays.

Introducing Sedimentary Rocks

1. Say that sedimentary rocks will be explored in this session. Define sediment.
2. Ask how rocks might be broken down into fragments.
3. Explain processes of weathering and erosion.

Investigating Sediments

1. Students work in groups to examine three types of sediment with magnifying lenses.
2. Distribute materials. Ask questions to encourage observation, especially regarding differences in size and shape of grains.
3. Explain that the terms sand, silt, clay refer to grain size. Define coarse and fine.

Observing Sedimentary Layers

1. Have students pour all three sediment samples into cup.
2. Present brief scenario of river transport and deposition.
3. Have students add water to cups, with all students having chance to stir, then cups placed in middle of area for all to see. Cups should not be moved again.
4. Students observe settling process. Class discusses their results. Ask which layer seems to have settled to the bottom and why.
5. Emphasize that **layered** texture is one of the most important properties of sedimentary rocks. Ask how this layering might happen. [Sediment tends to settle out of water in layers, which can often still be seen when it dries.]

6. Grains of sedimentary rocks are usually rounded, not sharp edged. Discuss why.
7. Explain that sedimentary rocks harden when piles of sediment accumulate in layers, and are compacted. Water is squeezed out of spaces between grains, and minerals left behind cement the grains together.
8. Discuss landforms on the Earth where sedimentary rocks can be found.

Investigating Nearby Soil
1. Ask students which three sediments are in soil near school.
2. Take students outside to collect samples.
3. In class, add water to their samples.
4. Students stir samples as before and place near other sedimentary profile.
5. Have students scoop off any material floating on top.
6. Discuss results of this sedimentary profile and compare to earlier one.
7. Explain that all soils contain mainly sediments (sand, silt, and clay) plus some organic material (humus).

Session 6: Formation of Metamorphic Rocks

Getting Ready
1. Use newspaper to cover student work areas.
2. Assemble materials on trays.
3. Gather extra set of materials for teacher demonstration. **Do the modeling activity yourself at least once before presenting.**

Introducing Metamorphic Rocks
1. Tell students that third class of rocks is metamorphic.
2. Explain that metamorphic rocks form when rocks exposed to heat and/or pressure.
3. Explain that students will create a clay model to show how metamorphic rocks form.

The Formation of Metamorphic Rocks
1. Demonstrate how students should form 20 red and 20 blue clay spheres, make yellow clay into a pancake, and green clay into two pancakes.
2. Distribute materials.
3. Based on the narrative provided, tell the story of the ancient lake bed. At appropriate points have students manipulate their clay model to correspond to the changes caused by environmental and geological events in the narrative.
4. Collect the pieces of model metamorphic rocks formed for use in the next session.
5. Encourage discussion of students' observations. Ask students to describe the metamorphic layers.
6. Explain that metamorphic rocks usually have a sheet-like texture and are usually much harder than sedimentary rocks from which they are most often derived.
7. Distribute class set of samples and have students look at slate and shale.
Which is sedimentary? [shale, layers of clay and silt] Which metamorphic? [slate, made from shale subjected to great heat and pressure; it is harder and more difficult to break]

Session 7: Recycling the Earth's Crust

Getting Ready
1. Use newspaper to cover student work areas.
2. Assemble materials on trays, with metamorphic model from last session.
3. Gather extra set of materials for teacher demonstration. **Do the modeling activity yourself at least once before presenting.**

The Changing Crust
1. Explain that studies show Earth's crust always moving and changing. If appropriate, introduce basic idea of theory of plate tectonics.
2. Explain that rocks produced by one process are often acted on and changed by others—this session's clay model simulates how such changes can take place.

Recycling the Crust
1. Distribute trays of materials.
2. Ask what type of rock the model represents. [metamorphic]
3. Begin the geological narrative provided, adapting as needed, making sure students complete each stage of the modeling process before going on to the next step.

The Rock Cycle
1. Encourage a discussion of the processes in the narrative, emphasizing that the changes occur over long periods of time.
2. Clarify that the rock cycle takes place within strata, and that rocks often go through transformations throughout their long histories and become different kinds of rocks. Any such transformation is part of the rock cycle.

Session 8: Classifying Rocks and Minerals

Getting Ready
1. Assemble materials on trays.
2. Select a "mystery rock" for each group.
3. Duplicate student data sheets.
4. List classifications and names of rocks and minerals on chalkboard, but cover names with paper until later.

Classifying—Using What We Know
1. Conduct brief review if needed; encourage student questions and responses.
2. Distribute class sets of rock and mineral samples with keys, lenses, pencils, and data sheets.
3. Have students work carefully in groups to classify each sample.
4. When groups agree, they should fill in data sheet.
5. Have groups report on their classifications.
6. Lift paper on board to reveal geological classification.

Mystery Rocks

1. Have groups observe and attempt to classify a "mystery rock." Distribute the rocks and data sheet. Have students record conclusions and reasoning on data sheet.

2. When finished, students place rock on data sheet in center of table so all can see.

3. Ask students to walk around to look at work of others and check off whether or not they agree or disagree with the classifications.

4. Groups then reconvene to consider comments on their classification and decide on final opinion. Groups should choose a spokesperson.

5. Representative from each group holds up rock and reports on classification and reasoning. Those who disagree report their observations that led to different conclusion.

6. Emphasize that scientists often disagree. From controversy science grows and changes. Just like the Earth's crust, science too is always changing!

A Giant Burning Rock Rose Up

A Lakota (Sioux) creation story says that before life began, in the very beginning, everything existed in the mind of Wakan-Tanka. All of the things that were to be existed only as spirits. These spirits traveled around in space looking for a place to come into being. They traveled to the Sun, but it was too hot. When they finally came to the Earth, it was without life or land, covered with the great waters. Then, out of the waters, a giant burning rock rose up. This rock made the dry land appear. Clouds formed from the steam it created. This made it possible for all the plants and animals to come forth. The Lakota call this first rock Tunka-shila, or "Grandfather Rock," and the story is intended to call forth respect for this first rock and for all rocks. In the sweat lodge, they say, when water strikes the heated stones and the steamy mist rises, it brings back that very first moment of creation—that is why the people sing to Grandfather Rock.

This story is adapted from the highly recommended book, with superb stories and educational activities, entitled Keepers of the Earth: Native American Stories and Environmental Activities for Children by Michael J. Caduto and Joseph Bruchac, Fulcrum, Inc., Golden, Colorado, 1988. Your students may notice the geological insight of the story, in its representation of the key role of volcanic eruption in creating (and re-creating) the land masses of Earth, as well as the connection it makes between volcanic eruption and atmospheric conditions.

Rocks and Minerals Key

Rocks & Minerals Key
1. schist   6. galena
2. halite   7. slate
3. granite   8. obsidian
4. quartz   9. shale
5. basalt   10. conglomerate

Rocks & Minerals Key
1. schist   6. galena
2. halite   7. slate
3. granite   8. obsidian
4. quartz   9. shale
5. basalt   10. conglomerate

Rocks & Minerals Key
1. schist   6. galena
2. halite   7. slate
3. granite   8. obsidian
4. quartz   9. shale
5. basalt   10. conglomerate

Rocks & Minerals Key
1. schist   6. galena
2. halite   7. slate
3. granite   8. obsidian
4. quartz   9. shale
5. basalt   10. conglomerate

Rocks & Minerals Key
1. schist   6. galena
2. halite   7. slate
3. granite   8. obsidian
4. quartz   9. shale
5. basalt   10. conglomerate

Rocks & Minerals Key
1. schist   6. galena
2. halite   7. slate
3. granite   8. obsidian
4. quartz   9. shale
5. basalt   10. conglomerate

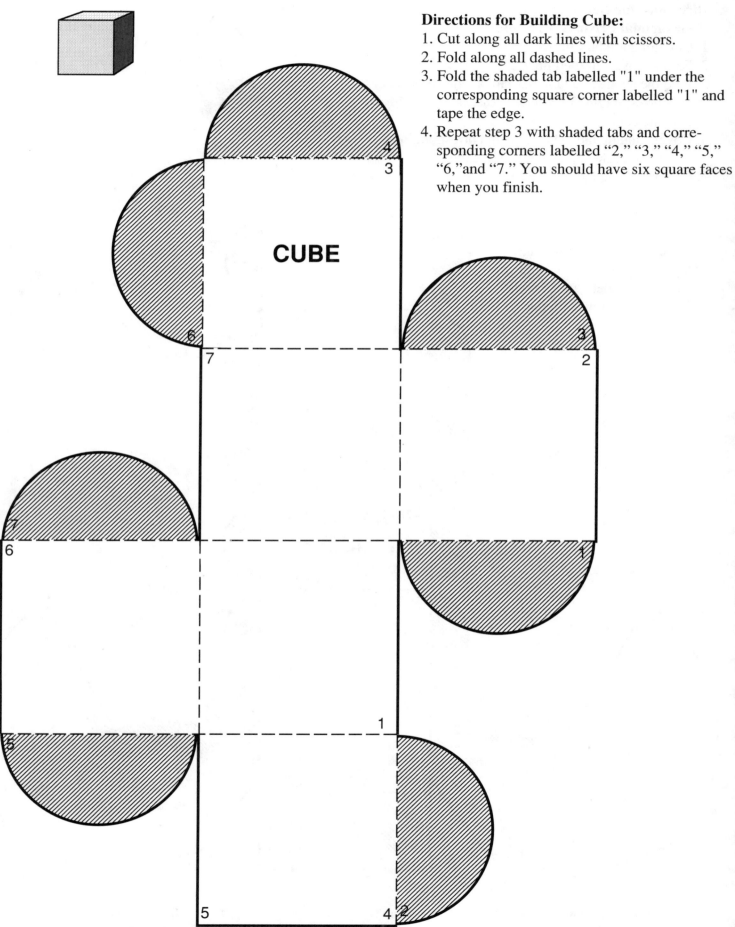

Directions for Building Cube:
1. Cut along all dark lines with scissors.
2. Fold along all dashed lines.
3. Fold the shaded tab labelled "1" under the corresponding square corner labelled "1" and tape the edge.
4. Repeat step 3 with shaded tabs and corresponding corners labelled "2," "3," "4," "5," "6,"and "7." You should have six square faces when you finish.

CUBE

LHS GEMS: *Stories in Stone*

Directions for Building
Hexagonal Prism & Pyramid:

1. Cut along all dark lines with scissors.
2. Fold along all dashed lines.
3. Fold the shaded rectangular tab labelled "1" under the corresponding rectangle corner labelled "1" and tape the edge.
4. Cover triangle labelled "2" with corresponding shaded triangle "2," and tape the edge.
5. Cover triangle labelled "3" with corresponding shaded triangle "3," and tape the edge.
6. Tuck semicircular tabs under corresponding triangles and tape the edges. You should have a six-sided "tube" with six-sided pyramids on the ends.

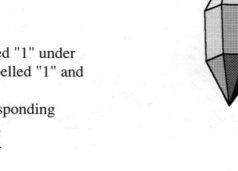

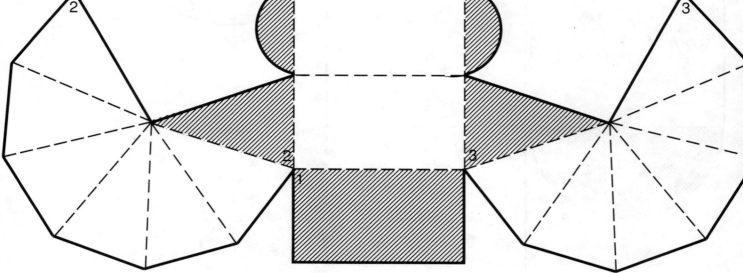

HEXAGONAL PRISM & PYRAMID

Name_____   Date_____

Observing Crystal Formation

Crystal Formation at Room Temperature

1. Place a very small amount (less than 1/8 teaspoon) of salol on a metal spoon.

2. Melt the salol by holding the spoon more than an inch (3 cm) above the flame.

3. Remove the spoon from the flame.

4. Add a few grains of salol as "seed crystals."

5. Prop up the handle so the spoon stays level.

6. Look at the crystals with a magnifying lens, and draw what you see.

Crystals at Room Temperature

Crystal Formation at Low Temperature

7. Remelt the crystals in **one** of the spoons.

8. Rest the bowl of this spoon on an ice cube.

9. Draw the shapes of the crystals that result when the salol cooled at a low temperature. Use the magnifying lens to compare the crystals at both temperatures.

Crystals at Low Temperature

10. Describe how the shapes and sizes of the crystals differ when they cooled at room temperature and at a low temperature.

Name_____   Date_____

Observing Crystal Formation

With a magnifying lens, observe the crystals formed at room temperature and draw what you see below:

With a magnifying lens, observe the crystals formed at a cooler temperature and draw what you see below:

Describe how the shapes and sizes of the crystals were different when they cooled at room temperature and at a lower temperature.

LHS GEMS: *Stories in Stone*

Rock Type
Description and Classification

Minerals

What It Is
Naturally occurring solids made of a single substance often found in crystal shapes.

Look for
Uniform appearance, crystal shapes, only one type of grain.

Igneous Rock

What It Is
Formed by cooling of magma. Slow cooling forms large crystals. Fast cooling forms tiny crystals or no crystals.

Look for
Interlocking grains with angular shapes.

Sedimentary Rock

Look for
Rounded grains, often layered. Chips off easily.

Metamorphic Rock

What It Is
Previously formed rocks undergo intense heat and pressure making them more compact and banded.

Look for
Sheet-like texture, flattened minerals in bands. Rock is very hard.

Observation and Display of "Mystery Rock"

Name of Our Group

Observations

[PLACE SAMPLE HERE]

Other Geologists' Opinions (√):

☐☐☐☐☐☐☐☐☐☐ **Mineral**
☐☐☐☐☐☐☐☐☐☐ **Igneous**
☐☐☐☐☐☐☐☐☐☐ **Sedimentary**
☐☐☐☐☐☐☐☐☐☐ **Metamorphic**

**What type of sample is this?
Here is our group's opinion:**

(circle one)

Igneous

Sedimentary

Metamorphic

Mineral

TEACHER'S GUIDES

Acid Rain
Grades 6–10

Animal Defenses
Preschool–K

Animals in Action
Grades 5–9

Bubble Festival
Grades K–6

Bubble-ology
Grades 5–9

Build It! Festival
Grades K–6

Buzzing A Hive
Grades K–3

Chemical Reactions
Grades 6–10

Color Analyzers
Grades 5–9

Convection: A Current Event
Grades 6–9

Crime Lab Chemistry
Grades 4–8

Discovering Density
Grades 6–10

Earth, Moon, and Stars
Grades 5–9

Earthworms
Grades 6–10

Experimenting with Model Rockets
Grades 6–10

Fingerprinting
Grades 4–8

Frog Math: Predict, Ponder, Play
Grades K–3

Global Warming & the Greenhouse Effect
Grades 7–10

Group Solutions
Grades K–4

Height-O-Meters
Grades 6–10

Hide A Butterfly
Preschool–K

Hot Water and Warm Homes from Sunlight
Grades 4–8

In All Probability
Grades 3–6

Investigating Artifacts
Grades K–6

Involving Dissolving
Grades 1–3

Ladybugs
Preschool–1

Liquid Explorations
Grades 1–3

Mapping Animal Movements
Grades 5–9

Mapping Fish Habitats
Grades 6–10

Math Around the World
Grades 5–8

Moons of Jupiter
Grades 4–9

More Than Magnifiers
Grades 6–9

Mystery Festival
Grades 2–8

Of Cabbages and Chemistry
Grades 4–8

Oobleck: What Do Scientists Do?
Grades 4–8

Paper Towel Testing
Grades 5–9

Penguins And Their Young
Preschool–1

QUADICE
Grades 4–8

River Cutters
Grades 6–9

Stories in Stone
Grades 4–9

Terrarium Habitats
Grades K–6

Tree Homes
Preschool–1

Vitamin C Testing
Grades 4–8

ASSEMBLY PRESENTER'S GUIDES

The "Magic" of Electricity
Grades 3–6

Solids, Liquids, and Gases
Grades 3–6

EXHIBIT GUIDES

Shapes, Loops & Images
All ages

The Wizard's Lab
All ages

HANDBOOKS

GEMS Teacher's Handbook

GEMS Leader's Handbook

Once Upon A GEMS Guide
(Literature Connections)

Insights & Outcomes
(Assessment)

A Parent's Guide to GEMS

To Build A House
(Thematic Approach to Teaching Science)

Write, call, or Fax:

GEMS
Lawrence Hall of Science
University of California
Berkeley, CA 94720-5200

(510) 642-7771

fax: (510) 643-0309

155

More on Themes

The word "themes" is used in many different ways in both ordinary usage and in educational circles. In the GEMS series, themes are seen as key recurring ideas that cut across all the scientific disciplines. Themes are bigger than facts, concepts, or theories. They link various theories from many disciplines. They have also been described as "the sap that runs through the curriculum," to convey the sense that they permeate through and arise from the curriculum. By listing the themes that run through a particular GEMS unit on the title page, we hope to assist you in seeing where the unit fits into the "big picture" of science, and how the unit connects to other GEMS units. The theme "Patterns of Change," for example, suggests that the unit or some important part of it exemplifies larger scientific ideas about why, how, and in what ways change takes place, whether it be a chemical reaction or a caterpillar becoming a butterfly. GEMS has selected 10 major themes:

Systems & Interactions	**Scale**
Models & Simulations	**Structure**
Stability	**Energy**
Patterns of Change	**Matter**
Evolution	**Diversity & Unity**

If you are interested in thinking more about themes and the thematic approach to teaching and constructing curriculum, you may wish to obtain a copy of our handbook, *To Build A House: GEMS and the Thematic Approach to Teaching Science.* For more information and a catalog, write or call GEMS, Lawrence Hall of Science, University of California, Berkeley, CA 94720. (510) 642-7771. **Thanks for your interest in GEMS!**

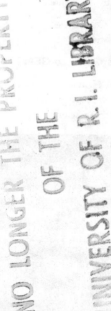

LHS GEMS